SpringerBriefs in Probability and Mathematical Statistics

Editor-in-chief

Mark Podolskij, Aarhus C, Denmark

Series editors

Nina Gantert, Münster, Germany
Richard Nickl, Cambridge, UK
Sandrine Péché, Paris, France
Gesine Reinert, Oxford, UK
Mathieu Rosenbaum, Paris, France
Wei Biao Wu, Chicago, USA

More information about this series at http://www.springer.com/series/14353

Tadahisa Funaki

Lectures on Random
Interfaces

 Springer

Tadahisa Funaki
Graduate School of Mathematical Sciences
The University of Tokyo
Tokyo, Japan

ISSN 2365-4333 ISSN 2365-4341 (electronic)
SpringerBriefs in Probability and Mathematical Statistics
ISBN 978-981-10-0848-1 ISBN 978-981-10-0849-8 (eBook)
DOI 10.1007/978-981-10-0849-8

Library of Congress Control Number: 2016959388

Printed on acid-free paper

This Springer imprint is published by Springer Nature
The registered company is Springer Nature Singapore Pte Ltd.
The registered company address is: 152 Beach Road, #22-06/08 Gateway East, Singapore 189721, Singapore

To Noriko

Preface

The aim of this book is to discuss random interfaces in several of their aspects. In the case where a phase coexistence occurs, interfaces are created to separate distinct phases. We study randomly fluctuating interfaces in several different settings and from different points of view: discrete/continuous, microscopic/macroscopic, or static/dynamic. Especially, the following four topics are covered:

1. Scaling limits for the $\nabla\varphi$-interface model with pinning
2. Dynamic Young diagrams
3. Sharp interface limits for a stochastic Allen-Cahn equation
4. The Kardar-Parisi-Zhang (KPZ) equation

Assuming that the interface is represented as a height function measured from a fixed reference discretized hyperplane, the system is governed by the Hamiltonian of gradients of the height functions. This is a kind of effective interface model, called the $\nabla\varphi$-interface model. The scaling limits leading from the microscopic level to the macroscopic one are studied for Gaussian or non-Gaussian random interfaces with a pinning effect in the case where the rate functional of the corresponding large deviation principle has nonunique minimizers. The analytic theory for studying the minimizers, which are characterized as solutions of a variational problem, is well developed, and it is known that the minimizers satisfy certain elliptic partial differential equations with free boundary conditions. This part reports joint work with E. Bolthausen and others [31, 32, 111], see also [103].

Young diagrams determine decreasing interfaces, and we consider their dynamics mostly in two-dimensional case. The large-scale behavior of such dynamics is studied from the viewpoints of the hydrodynamic limit, the non-equilibrium fluctuation theory, and the large deviation principle. Vershik curves are derived in the limit as unique stationary solutions of the hydrodynamic equations. We will refer to related static results on two- and three-dimensional random Young diagrams. This part is based on joint work with M. Sasada and others [114, 115].

Sharp interface limits for the Allen-Cahn equation, which is a reaction-diffusion equation with a balanced bistable reaction term, lead to the motion by mean curvature for the interfaces. Its stochastic perturbation, the stochastic Allen-Cahn equation, sometimes called the time-dependent Ginzburg-Landau model, a stochastic quantization, or a dynamic $P(\phi)$-model is studied. A brief introduction to the space-time Gaussian white noise, colored noise, Q-Brownian motion, and stochastic integrals is given. A regularity property of solutions of stochastic partial differential equations (SPDEs) of parabolic type with additive noises is also discussed. The references for this part are Funaki [96–98], Funaki and Yokoyama [118], and Weber [216]. A survey is given in [100]; see also [105].

The Kardar-Parisi-Zhang (KPZ) equation describes a growing interface with fluctuation and recently attracts a lot of attentions. It is an ill-posed SPDE and requires a renormalization. In particular, its invariant measures are studied. This part is based on joint work with J. Quastel [112] and [107]. Both the random Young diagrams and the KPZ equation have aspects of "integrable probability," and this plays a fundamental role in recent developments. This book discusses from another important aspect of "stochastic analysis," and these two aspects supplement each other.

My goal was to explain the motivations behind the problems and the ideas of the proofs rather than giving full details. Sometimes our arguments are intuitive or rather heuristic, but I believe this will facilitate readers' quick understanding. Instead, I have tried to give relevant references. Some conjectures are also mentioned. The common theme throughout the book is random interfaces, and the subjects discussed in different chapters are linked. However, each chapter can be read independently.

It is the time for me to thank many people to whom I am deeply indebted. Over long years, I was stimulated particularly by Erwin Bolthausen and Herbert Spohn, and I did learn a lot from them. Some of the main ingredients of this book are outcomes of fruitful discussions and collaborations with both of them.

The book grew out of notes on lectures given at Kyushu University in 2010 (Chap. 2), Beijing University in 2012 (Chaps. 3, 4, and 5), Waseda University in 2012 and 2013 (Chaps. 3 and 4), University of Tokyo in 2014 and 2016 (Chaps. 1, 2, 3, 4, and 5), the Korean Summer School in 2015 (Chaps. 1, 3, 4, and 5), and the UK-Japan Winter School in 2016 (Chaps. 3, 4, and 5). I thank Hirofumi Osada of Kyushu University; Dayue Chen and Yong Liu of Beijing University; Yoshihiro Shibata of Waseda University; Kyeonghun Kim, Panki Kim, and Hyun-Jae Yoo, the organizers of the Third NIMS Summer School in Probability 2015 held at NIMS, Daejeon, Korea; and Martins Bruveris, Darryl Holm, Tudor Ratiu, and Hiroaki Yoshimura, the organizers of the UK-Japan Winter School "Classic and Stochastic Geometric Mechanics" held at Imperial College, London, for their kind invitations.

I thank the anonymous referees for their useful comments. I am grateful to Ludovic Goudenège, Vincent Pit (Figs. 1.1 and 1.7), Makiko Sasada (Figs. 2.1, 2.2, 2.4, and 2.11), and Kazumasa Takeuchi (Fig. 5.1) for providing figures with kind permissions. I also thank Masayuki Nakamura and Catriona Byrne of Springer-Verlag for their kind encouragement and patience. The editorial office helped to draw the many figures used in the book and to improve the English writing.

The research leading to the results presented in this book was partially supported by JSPS KAKENHI Grant Numbers (A) 18204007, 21654021, (A) 22244007, (B) 26287014, and 26610019.

Tokyo, Japan   Tadahisa Funaki
August 2016

Contents

Chapter 1
Scaling Limits for Pinned Gaussian Random Interfaces in the Presence of Two Possible Candidates

The macroscopic shape of crystals is usually described by variational problems. We first explain those characterizing the Wulff shape, the Winterbottom shape and also those with two media, with a pinning effect. We give examples of minimizers in a pinning case. Then, we explain underlying microscopic models such as the Ising model and the $\nabla\varphi$-interface model. Macroscopic variational problems and microscopic models are linked by a large deviation principle, or a law of large numbers. We will focus on the $\nabla\varphi$-interface model with a pinning. For such model, results for $d = 1, n \geq 1$ (obtained with Bolthausen and Otobe [32, 111]) and those for $d \geq 3, n = 1$ (obtained with Bolthausen and Chiyonobu [31]) will be presented, where d is the dimension of the base space, while n is the dimension of the value (target) space. See [103, 116] for the $\nabla\varphi$-interface model.

1.1 Macroscopic Variational Problems

1.1.1 Wulff Shape and Winterbottom Shape

The shape of a crystal (or a droplet of water located in vapor) is determined so as to minimize the total surface tension. Let a direction-dependent (anisotropic) surface tension $\sigma : S^{d-1} \to (0, \infty)$ be given, where $S^{d-1} = \{\vec{n} \in \mathbb{R}^d; |\vec{n}| = 1\}$. Then, for a domain V (describing a crystal or a droplet) in $\mathbb{R}^d$ with a boundary $S = \partial V$, we define the *total surface tension* of S (or a total energy of the interface S) by

$$J(S) = \int_S \sigma(\vec{n}(x))dx,$$

© The Author(s) 2016
T. Funaki, *Lectures on Random Interfaces*, SpringerBriefs in Probability and Mathematical Statistics, DOI 10.1007/978-981-10-0849-8_1

where $\vec{n}(x)$ is the outward normal vector at $x \in S$ and dx is the surface element of S. The *Wulff shape* (introduced in 1901, see [220]) is determined as the minimizer of the variational problem

$$\min_{\text{vol } V=v} J(S),$$

for each given volume $v > 0$.

When the droplet is placed on a solid substrate, in addition to the water-vapor surface tension σ, one needs to consider the effect of the water-solid surface tension σ_{WS}. The equilibrium shape of a droplet is called the *Winterbottom shape* (introduced in 1967). In the context of the $\nabla\varphi$-interface model discussed later, if the solid substrate is located at the height level 0, the variational problem is formulated with a pinning effect (see Sect. 1.1.3) and considered under the conditions $h \geq 0$ and the total volume $\int h dx = v > 0$, where h describes the height of the interface, see Bolthausen and Ioffe [33].

1.1.2 Variational Problem with Two Media

We now consider the situation that the interface S can be described by a height function h measured from a fixed reference hyperplane $\{(x,0); x \in D\} \subset \mathbb{R}^{d+1}$ such that $S = \{(x,y); y = h(x), x \in D\}$, namely, S is represented as the graph of $y = h(x)$. Such a model is called an effective model.

Let $D \subset \mathbb{R}^d$ be a bounded region and for $h : D \to \mathbb{R}$ (or $\mathbb{R}^n$ later), consider the total energy given by

$$J(h) = \int_D \{\sigma(\nabla h(x)) - Q(x)1_{\{h(x)\leq 0\}}\}dx, \tag{1.1}$$

where $\sigma : \mathbb{R}^d \to \mathbb{R}$ is a convex function (e.g., $\sigma(u) = \frac{1}{2}|u|^2$) and $Q(x) \geq 0$. If h takes negative values, $J(h)$ becomes smaller and this means that the negative side has an advantage. In other words, we consider a domain $D \times \mathbb{R}$, which is filled with a predominant medium in the lower half $D \times (-\infty, 0]$ of this domain and a minor medium in the upper half $D \times (0, \infty)$. The analytic theory for J was developed by Alt and Caffarelli [8], Alt, Caffarelli, and Friedman [9], Weiss [218] and others.

One can show that the minimizer h of J (Fig. 1.1) satisfies the Euler equation

$$\text{div}\{\nabla\sigma(\nabla h)\} \equiv \sum_{i=1}^d \frac{\partial}{\partial x_i}\left(\frac{\partial\sigma}{\partial u_i}(\nabla h)\right) = 0 \quad \text{on } D \setminus \Gamma, \tag{1.2}$$

where $\Gamma = \{x \in D; h(x) = 0\}$, and the free boundary condition

$$\Psi(\nabla h^+) - \Psi(\nabla h^-) = Q \quad \text{on } \Gamma, \tag{1.3}$$

 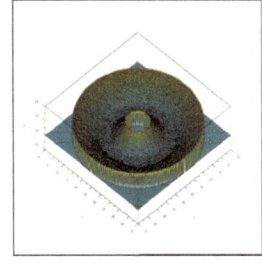 

$D = \text{disc}, \ h|_{\partial D} \geq 0$ $\qquad\qquad D = \text{annulus}$

Fig. 1.1 Shapes of minimizers

where $\Psi(u) = u \cdot \nabla\sigma(u) - \sigma(u)$ and $\nabla h^{\pm}$ are the gradients of h from the positive and negative sides, respectively. In fact, to derive (1.2), let us consider $J(h)$ ignoring the jumps at Γ so that we may examine $J(h) = \int_D \sigma(\nabla h(x))dx$ or $\int_D \{\sigma(\nabla h(x)) - Q(x)\}dx$. Then, for every test function $\varphi \in C_0^{\infty}(D)$, we have

$$\frac{d}{d\varepsilon}J(h + \varepsilon\varphi)\Big|_{\varepsilon=0} = \int_D \frac{d}{d\varepsilon}\sigma(\nabla h(x) + \varepsilon\nabla\varphi(x))\Big|_{\varepsilon=0} dx$$

$$= \int_D (\nabla\sigma)(\nabla h) \cdot \nabla\varphi dx$$

$$= -\int_D \text{div}\left((\nabla\sigma)(\nabla h)\right)\varphi dx.$$

This implies (1.2) for the minimizer h of J. See Friedrichs [86] for the derivation of (1.3) in a general case.

In particular, when $\sigma(u) = \frac{1}{2}|u|^2$, the Euler equation and the free boundary condition, sometimes called the *Young's relation*, are given by

$$\Delta h = 0 \quad \text{and} \quad |\nabla h^+|^2 - |\nabla h^-|^2 = 2Q(x),$$

respectively. We give a derivation of the Young's relation (1.3) in this simple case assuming $D = [0, 1]$ and $Q(x) \equiv Q$ in addition to $\sigma(u) = \frac{1}{2}|u|^2$, under the boundary conditions $h(0) = a, h(1) = b$. From (1.2), that is $h'' = 0$ on $D \setminus \Gamma$, the possible minimizers must have the form in Figs. 1.2 or 1.3, if they touch the height level 0. Let us denote $P_1 = (x_1, 0)$ and $P_2 = (1 - x_2, 0)$ in Fig. 1.2. Then we have

$$J = \frac{1}{2}\left(\frac{a}{x_1}\right)^2 x_1 + \frac{1}{2}\left(\frac{b}{x_2}\right)^2 x_2 - Q(1 - x_1 - x_2)$$

$$= \frac{a^2}{2x_1} + \frac{b^2}{2x_2} - Q(1 - x_1 - x_2).$$

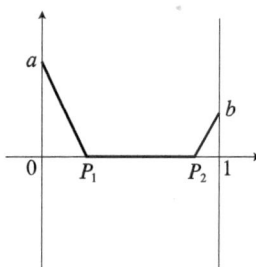

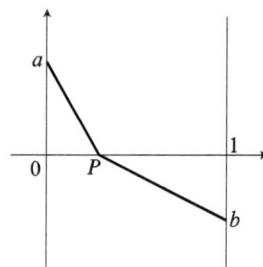

Fig. 1.2 The case $a, b > 0$ **Fig. 1.3** The case $a > 0, b < 0$

Therefore, for the minimizers it holds that

$$\frac{\partial J}{\partial x_1} = -\frac{a^2}{2x_1^2} + Q = 0,$$

and this implies

$$\left(\frac{a}{x_1}\right)^2 (= |\nabla h^+|^2) = 2Q,$$

or, $\tan \alpha_1 = \sqrt{2Q}$ if we denote the refraction angle at P_1 by α_1.

1.1.3 Variational Problem with a Pinning Effect

Section 1.1.2 discussed the situation where the negative side has an advantage. Here we consider the situation that a single point $\{0\}$ (or $D \times \{0\}$) has an advantage. This is called a pinning effect. The problem becomes a bit singular compared with that discussed in the preceding subsection.

The energy of the height of an interface $h : D \to \mathbb{R}$ (or $\mathbb{R}^n$) with a pinning at $0 \in \mathbb{R}$ (or $\mathbb{R}^n$) is given by

$$J(h) = \int_D \{\sigma(\nabla h(x)) - \xi 1_{\{h(x)=0\}}\}dx, \quad \xi > 0. \tag{1.4}$$

The Euler equation and the free boundary condition for the minimizer h are similar to the ones for the two-media problem, at least if $n = 1$ and $h|_{\partial D} \geq 0$.

1.1.4   Examples of Minimizers of J with a Pinning Effect

Here, we consider the simple case: $d = 1$, $n \geq 1$, $\sigma = \frac{1}{2}|u|^2$, $D = [0, 1]$. In this case, for $h : [0, 1] \to \mathbb{R}^n$ satisfying $h(0) = a$, $h(1) = b$, the energy J with a pinning at 0 is given by

$$J(h) = \frac{1}{2} \int_0^1 |\dot{h}(x)|^2 \, dx - \xi \big| \{x \in [0, 1]; \, h(x) = 0\} \big|,$$

where $|\{\cdots\}|$ denotes the Lebesgue measure.

This energy J has two possible minimizer candidates $\bar{h}$ and $\hat{h}$, where $\bar{h}$ is a straight line connecting a and b, see Fig. 1.4. The Euler equation is given by $\Delta h = 0$. This means that the minimizer should be a line except when it touches 0. In $\hat{h}$, let us denote $P_1 = (x_1, 0)$ and $P_2 = (1 - x_2, 0)$. Then, the free boundary condition, called Young's relation, is

$$|a|/x_1 = |b|/x_2 = \sqrt{2\xi},$$

as we discussed in Sect. 1.1.2.

It depends on the boundary values a and b which of $\bar{h}$ or $\hat{h}$ is the real minimizer. Indeed, the minimizer is determined by the balance between two conflicting effects produced by the two terms of J. If a and b are far from 0, then it is better to stay on the straight line connecting a and b, so that $\bar{h}$ is the minimizer. On the other hand, if a and b are close to 0, then, even though one pays a price for leaving $\bar{h}$, it is better to go to the level 0 and stay there for a while, so that $\hat{h}$ becomes the minimizer.

If a and b satisfy a certain balance condition, both $\bar{h}$ and $\hat{h}$ can be minimizers at the same time, i.e., $J(\bar{h}) = J(\hat{h})$ holds. We call such a case critical (or coexistence). In particular, when $d = n = 1$, it is known that the condition for the coexistence is given by $\sqrt{|a|} + \sqrt{|b|} = (2\xi)^{1/4}$, $ab > 0$, see [32].

The question we will ask is the following: which minimizer really appears in the critical case. Macroscopically, the two minimizers above have the same degree of

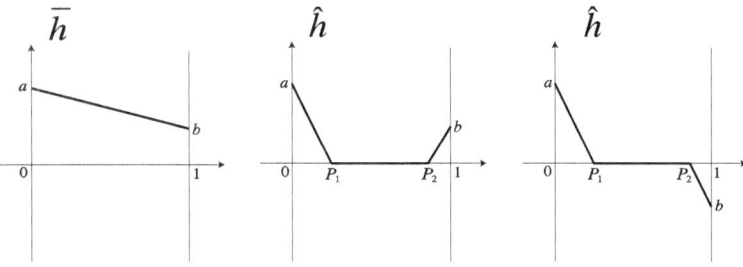

Fig. 1.4 Possible candidates of minimizers

importance and we can not distinguish them in terms of J only. However, we can
determine the difference if we examine the system from the microscopic level.

1.2 Microscopic Models

1.2.1 Wulff Shape from Microscopic Models

Dobrushin, Kotecký, and Shlosman [65], Ioffe and coauthor [154–156] proved the
large deviation principle (LDP) for the Ising model and derived the Wulff shape
in a scaling limit. The Ising model is a microscopic model, while the Wulff shape
is a macroscopic shape of a crystal. Several related results are known for the so-
called $\nabla\varphi$-interface model taken as a microscopic model. Bolthausen and Ioffe [33]
studied the $\nabla\varphi$-interface model with a wall and a pinning effect when $d = 2$ and
derived the Winterbottom shape as a macroscopic shape of a crystal. Deuschel,
Giacomin, and Ioffe [63] proved the LDP for the $\nabla\varphi$-interface model and derived the
Wulff shape in the scaling limit in this setting. Funaki and Sakagawa [113] studied
the $\nabla\varphi$-interface model with a weak self potential, proved the LDP with $J(h)$ given
in (1.1) as an (unnormalized) rate function, and derived the free boundary problem
discussed in Sect. 1.1.2. In the following, we explain these results in slightly more
details.

We consider the 2D or 3D *Ising model*. Let $D_N := [-N, N]^d \cap \mathbb{Z}^d, d = 2$ or 3, be
a large box and for a configuration $s = \{s_i; i \in D_N\} \in \{+1, -1\}^{D_N}$ on D_N define a
Hamiltonian $H_N(s)$ as the sum over all bonds $\langle i, j \rangle$ in D_N (i.e., $i, j \in D_N : |i-j| = 1$):

$$H_N(s) = -\frac{1}{2} \sum_{\langle i,j \rangle \subset D_N} s_i s_j + \sum_{i \in \partial^+ D_N} s_i.$$

The second sum means that we impose the minus boundary conditions $s_i = -1$ for
$i \in \partial^+ D_N := \{i \in \mathbb{Z}^d \setminus D_N; |i - j| = 1$ for some $j \in D_N\}$. The (grand canonical)
Gibbs measure with an inverse temperature $\beta > 0$ is defined as

$$P_{N,\beta}^-(s) = \frac{1}{Z_{N,\beta}^-} e^{-\beta H(s)}, \quad s \in \{+1, -1\}^{D_N},$$

where $Z_{N,\beta}^-$ is the normalization constant. It is known that the limit $m^* = m^*(\beta) :=$
$-\lim_{N \to \infty} \frac{1}{|D_N|} \sum_{i \in D_N} \langle s_i \rangle_{P_{N,\beta}^-}$ exists for every $\beta > 0$ and there exists a critical $\beta_c >$
0 such that a spontaneous magnetization, that is $m^*(\beta) > 0$, appears if $\beta > \beta_c$,
where $\langle \cdot \rangle_{P_{N,\beta}^-}$ denotes the mean under $P_{N,\beta}^-$ and $|D_N| = (2N + 1)^d$. For m with
$|m| < m^*$, define the canonical Gibbs measure as the conditional probability:

$$P_{N,\beta}^{m,-}(\cdot) = P_{N,\beta}^{-}\left(\cdot \left| \frac{1}{|D_N|} \sum_{i \in D_N} s_i = m \right.\right).$$

(We assume $m \cdot |D_N|$ is an integer for simplicity.) Let $\Gamma_N := \{i \in D_N; s_i = +1\}$. Then, it is known that the macroscopically scaled shape $\frac{1}{N}\Gamma_N$ (up to translation) converges under $P_{N,\beta}^{m,-}$ as $N \to \infty$ to the so-called *Wulff shape*, which is characterized as the minimizer of the variational problem

$$\min_{\Gamma \subset D \,:\, \mathrm{vol}(\Gamma)=m} J(\partial\Gamma),$$

where $D = [-1, 1]^d$, $J(\partial\Gamma)$ is the *total surface tension*, given by

$$J(\partial\Gamma) = \int_{\partial\Gamma} \tau(\overrightarrow{n}(x))dx,$$

and $\tau(\overrightarrow{n})$ is the surface tension of the surface with a tilt orthogonal to the vector $\overrightarrow{n} \in S^{d-1}$, defined by

$$\tau(\overrightarrow{n}) = \lim_{N \to \infty} \frac{1}{(2N+1)^{d-1}} \log \frac{Z_N^{+,-}}{Z_N^{-}},$$

where $Z_N^{+,-}$ is the partition function (normalization constant) with the boundary condition indicated in the right most figure in Fig. 1.5 insisting a tilt perpendicular to $\overrightarrow{n}$, while Z_N^{-} is the partition function with the pure $-$ boundary condition. This can be shown through the LDP for $P_{N,\beta}^{-}$, which implies the LDP for

$$P_{N,\beta}^{m}(\cdot) = P_{N,\beta}^{-}\left(\cdot \left| \frac{1}{|D_N|} \sum_{i \in D_N} s_i \geq m \right.\right).$$

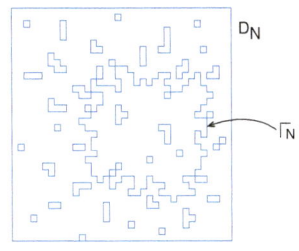 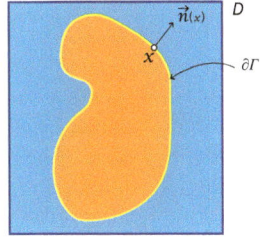 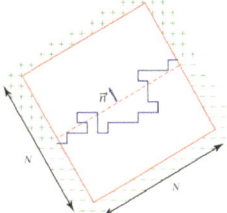

Fig. 1.5 Γ_N, $J(\partial\Gamma)$, $\tau(\overrightarrow{n})$

The 2D case was studied by Dobrushin, Kotecký, Shlosman, Ioffe and coauthor mentioned above, while the 3D case was studied by Bodineau [29], Cerf and Pisztora [46], and others.

As we mentioned above, the macroscopic crystal shapes are characterized by variational problems and the *large deviation principle* (LDP) connects the microscopic models with the macroscopic variational problems. We quickly recall this procedure: Let $\{\mu_N\}_N$ be a sequence of probability measures describing the microscopic model. Let h^N be the scaled heights (random variables) defined under μ_N. We say that an LDP holds for h^N with a rate function $J^*(h) \left(= J(h) - \inf J\right) \geq 0$ and a speed N if

$$\mu_N(h^N \sim h) \sim e^{-NJ^*(h)} \quad \text{as } N \longrightarrow \infty,$$

for each given h. This is a rather rough statement. The left-hand side is the probability that h^N is close to a given h (with respect to a certain distance or a topology), and this probability behaves asymptotically as the right-hand side. Under the LDP, we easily have the following concentration properties: Set $\mathcal{M} = \{h^*; \text{minimizers of } J\}$, then for any $\delta > 0$, there exists $c = c_\delta > 0$ such that

$$\mu_N\left(\text{dist}(h^N, \mathcal{M}) \geq \delta\right) \leq e^{-cN}.$$

As we already stated, our goal is to understand what happens for the microscopic model described by μ_N if the variational problem for J^* (and that for J) has several minimizers.

1.2.2   The $\nabla\varphi$-Interface Model with a Pinning

Let us first recall the $\nabla\varphi$-interface model, see [103] for more details. Let $D \subset \mathbb{R}^d$ be a given (bounded) macroscopic domain. For a microscopic domain $D_N = ND \cap \mathbb{Z}^d$ associated with D and microscopic height variables $\phi = (\phi_i)_{i \in D_N} \in \mathbb{R}^{D_N}$, define the Hamiltonian

$$H_N(\phi) = \sum_{\langle i,j \rangle \subset D_N} V(\phi_i - \phi_j),$$

with some boundary conditions at $\partial D_N := \{i \in D_N; |i - j| = 1 \text{ for some } j \in \partial^+ D_N\}$, where $V : \mathbb{R} \to \mathbb{R}$ is a given potential. Then, the Gibbs measure μ_N on $\mathbb{R}^{D_N^\circ}$, where $D_N^\circ = D_N \setminus \partial D_N$, is defined by

$$\mu_N(d\phi) = \frac{1}{Z_N} e^{-H_N(\phi)} d\phi, \quad d\phi = \prod_{i \in D_N^\circ} d\phi_i,$$

where Z_N is a normalization constant.

We first state a result of Bolthausen and Ioffe [33]. They considered μ_N^ε, $\varepsilon > 0$, defined similarly to μ_N by replacing $d\phi$ with $\prod_{i \in D_N^\circ}(d\phi_i + \varepsilon \delta_0(d\phi_i))$ for $d = 2$, $D = [-1, 1]^2$, $D_N = [-N, N] \cap \mathbb{Z}^2$, $V(u) = \frac{1}{2}|u|^2$ with null boundary conditions at ∂D_N, and the law $\mu_N^{\varepsilon,+}$ conditioned μ_N^ε to be $\phi_i \geq 0$ for all $i \in D_N$, and introduced a further conditional law $\mu_{N,v}^{\varepsilon,+}$ of $\mu_N^{\varepsilon,+}$ under the assumption that $\frac{1}{N^2}\sum_{i \in D_N}\frac{1}{N}\phi_i \geq v$ for $v > 0$. Note that the last sum corresponds to the macroscopic volume $\int_D h^N(x)dx$ of the droplet described by h^N introduced below. The factor $\varepsilon\delta_0(d\phi_i)$ describes the pinning effect at 0, and the conditioning on $\phi_i \geq 0$ means a hard wall is placed at the height level 0. Then, they studied the scaling from the microscopic object $\phi = (\phi_i)_{i \in D_N}$ to the macroscopic height function $h^N = (h^N(x))_{x \in D}$ defined by $h^N(x) = \frac{1}{N}\phi_{[Nx]}, x \in D = [-1, 1]^2$, where $[Nx]$ denotes the integer part of Nx taken componentwise. The macroscopic energy of $h : D \to [0, \infty)$ is defined by

$$J(h) = \frac{1}{2}\int_D |\nabla h|^2 dx - \Delta_\varepsilon|\{x \in D; h(x) = 0\}|,$$

with some Δ_ε. It is known that $\Delta_\varepsilon > 0$ if $\varepsilon > 0$ is sufficiently large. The minimizers of J, called *Winterbottom shapes*, are given by $h_{WB}(x) = (1 - c|x|^2) \vee 0$ and their shifts with some $c = c_\varepsilon > 0$, see Fig. 1.6. Bolthausen and Ioffe [33] proved that

$$\lim_{N \to \infty} \mu_{N,v}^{\varepsilon,+}\left(\text{dist}_{L^1(D)}\{h^N, \mathcal{M}\} \geq \delta\right) = 0 \qquad (1.5)$$

for every $\delta > 0$, where $\mathcal{M}$ is the set of all possible shifts of h_{WB}, i.e., $\mathcal{M} = \{h_{WB}(\cdot + y); y \in \mathbb{R}^2 \text{ such that the support of } h_{WB}(\cdot + y) \subset D\}$. Figure 1.7 indicates the appearance of spikes at microscopic level, so that the L^1-distance in (1.5) cannot be improved (certainly this is not possible in the L^∞-distance.)

We next state the result of Deuschel, Giacomin, and Ioffe [63]. They considered μ_N for a general convex potential $V \in C^2(\mathbb{R})$ such that $0 < c_- \leq V'' \leq c_+ < \infty$ with some $c_\pm > 0$ and having null boundary conditions on $D_N = ND \cap \mathbb{Z}^d$ determined by a general domain D. They proved the LDP for h^N under μ_N with the (unnormalized) rate function $\int_D \sigma(\nabla h(x))dx$, in which

$$\sigma(u) = -\lim_{\ell \to \infty}\frac{1}{|\Lambda_\ell|}\log\frac{Z_\ell(u)}{Z_\ell(0)}, \qquad u \in \mathbb{R}^d,$$

where $Z_\ell(u)$ is the partition function (normalization constant) for μ_ℓ defined on $\Lambda_\ell = [-\ell, \ell]^d \cap \mathbb{Z}^d$ with the boundary condition $\psi_i = i \cdot u, i \in \partial \Lambda_\ell$. The existence of σ is shown in [116].

Fig. 1.6 Macroscopic Winterbottom shape (Taken from [33])

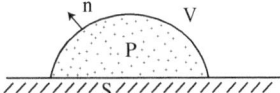

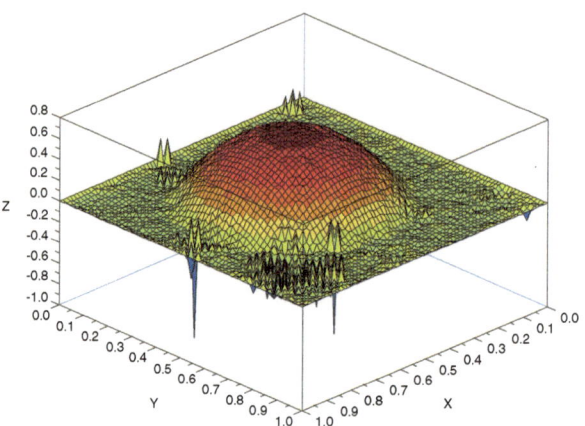

Fig. 1.7 Microscopic droplet: Winterbottom shape without hard wall condition

We finally state the result of Funaki and Sakagawa [113]. We considered the Hamiltonian

$$H_{N,U}(\phi) = \sum_{\langle i,j \rangle \subset D_N} V(\phi_i - \phi_j) + \sum_{i \in D_N} U(i/N, \phi_i),$$

with a self-potential U of the form $U(x, \phi) = Q(x)W(\phi)$. We proved, as a perturbation of the result of [63], the LDP for h^N under the Gibbs measure $\mu_{N,U}$ corresponding to the above Hamiltonian with the (unnormalized) rate function

$$\int_D \left\{ \sigma(\nabla h(x)) dx - AQ(x) 1_{\{h(x) \le 0\}} \right\} dx,$$

with $A = \alpha - \beta$, where $\alpha = \lim_{\phi \to \infty} W(\phi)$ and $\beta = \lim_{\phi \to -\infty} W(\phi)$, assuming $\alpha > \beta$. This functional is essentially the same as $J(h)$ in (1.1).

Our model: Let us introduce our microscopic model, called the $\nabla\phi$-interface model with a pinning under a special boundary condition. The system is defined on a d-dimensional square lattice cylinder of size N, being periodic in the 2nd to the dth coordinates:

$$D_N = \{0, 1, 2, \ldots, N\} \times \mathbb{T}_N^{d-1}, \quad \mathbb{T}_N^{d-1} = (\mathbb{Z}/N\mathbb{Z})^{d-1}.$$

In other words, we consider the d-dimensional box $\{0, 1, 2, \ldots, N\}^d$ and identify its boundaries except for the first coordinate, see Fig. 1.8. The reason for considering such a cylinder is that one can easily find macroscopic minimizers by extending the one-dimensional example given in Sect. 1.1.4, see Sect. 1.5 below. We denote $D_N^\circ = D_N \setminus \partial D_N$, $\partial D_N = \partial_L D_N \cup \partial_R D_N$, $\partial_L D_N = \{0\} \times \mathbb{T}_N^{d-1}$, and $\partial_R D_N = \{N\} \times \mathbb{T}_N^{d-1}$.

Fig. 1.8 Lattice cylinder

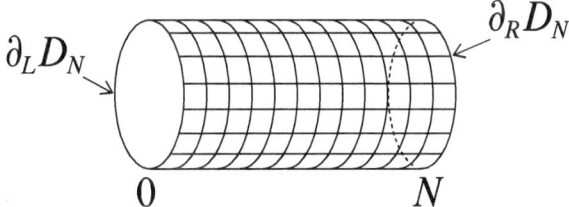

The microscopic objects (height functions when $n = 1$) $\phi = (\phi_i)_{i \in D_N} : D_N \to \mathbb{R}^n$ are fields defined on D_N. The energy (Hamiltonian) of ϕ is given as the sum over all bonds $\langle i, j \rangle$ (i.e., pairs of nearest neighbor sites) in D_N:

$$H_N(\phi) = \frac{1}{2} \sum_{\langle i,j \rangle \subset D_N} |\phi_i - \phi_j|^2, \qquad (1.6)$$

with the boundary conditions at ∂D_N determined from given macroscopic values $a, b \in \mathbb{R}^n$:

$$\phi_i = aN, \ i \in \partial_L D_N \quad \text{and} \quad \phi_i = bN, \ i \in \partial_R D_N. \qquad (1.7)$$

Then, the microscopic model for ϕ is described by the Gibbs measure with a pinning:

$$\mu_N^\varepsilon(d\phi) = \frac{1}{Z_N^\varepsilon} e^{-H_N(\phi)} \prod_{i \in D_N^\circ} \left(\varepsilon \delta_0(d\phi_i) + d\phi_i \right), \qquad (1.8)$$

where $\varepsilon \geq 0$ is a parameter called the strength of the pinning, and Z_N^ε is the normalization constant, called the partition function. Note that μ_N^0 (i.e. $\varepsilon = 0$) is Gaussian. When $d = 1$, μ_N^ε defines a random walk and we will consider the general non-Gaussian case in Sect. 1.3.

Scaling from the microscopic object ϕ to the macroscopic object h^N is defined as follows: Let $D = [0, 1] \times \mathbb{T}^{d-1}$ be a continuous cylinder of size 1, where $\mathbb{T}^{d-1} = (\mathbb{R}/\mathbb{Z})^{d-1}$. Then the macroscopic field $h^N = (h^N(x))_{x \in D} \in C(D, \mathbb{R}^n)$ is defined as

$$h^N \left(\frac{i}{N} \right) = \frac{1}{N} \phi_i, \quad i \in D_N, \qquad (1.9)$$

and its extension to D as a step function. We sometimes take a polilinear interpolation instead of step extension, see [31, 63]. Our problem is to find the limit of h^N under μ_N^ε as $N \to \infty$. A microscopic model corresponding to the two-media case was studied by [113], as we indicated above.

The effect of the microscopic pinning is macroscopically reflected in the *pinning free energy*, defined by

$$\xi^{\varepsilon} = \lim_{\ell \to \infty} \frac{1}{|\Lambda_{\ell}|} \log \frac{Z_{\Lambda_{\ell}}^{0,\varepsilon}}{Z_{\Lambda_{\ell}}^{0}}, \tag{1.10}$$

where $\Lambda_{\ell} = [1, \ell]^d \cap \mathbb{Z}^d$ (or one can take $[-\ell, \ell]^d \cap \mathbb{Z}^d$ as above), $|\Lambda_{\ell}| = \ell^d$, and $Z_{\Lambda_{\ell}}^{0,\varepsilon}, Z_{\Lambda_{\ell}}^{0}$ are the partition functions on Λ_{ℓ} with null boundary conditions with/without pinning, respectively.

It is known that there exists $\varepsilon_c \geq 0$ such that $\varepsilon > \varepsilon_c \iff \xi^{\varepsilon} > 0$ (i.e., localization) holds. Moreover, we have that

$d = 1, n \geq 3 \implies \varepsilon_c > 0$ (Pinning transition occurs),

$d \geq 1, n = 1$ or $d = 1, n = 2 \implies \varepsilon_c = 0$ (No transition occurs; always localized).

See Theorem 1.1 of [111] for $d = 1$ and Section 7 of [103] for localization when $d \geq 2, n = 1$. When $d = 1$, ϕ can be regarded as a random walk. In this case, if $n = 1$ or 2, ϕ is null-recurrent when $\varepsilon = 0$ and turns to be positive-recurrent for every $\varepsilon > 0$, so that ϕ is localized. On the other hand, if $n \geq 3$, ϕ is transient (i.e., delocalized) when $\varepsilon = 0$. This intuitively is the reason why $\varepsilon_c = 0$ when $d = 1, n = 1$ or 2, while $\varepsilon_c > 0$ when $d = 1, n \geq 3$.

1.3   Results for $d = 1$, $n \geq 1$

We take $D_N = \{0, 1, 2, \ldots, N\}$ and consider the microscopic system $\phi = \{\phi_i\}_{i \in D_N}$ distributed according to the measure μ_N^{ε} on $(\mathbb{R}^n)^{D_N^{\circ}}$ defined by

$$\mu_N^{\varepsilon}(d\phi) = \frac{1}{Z_N^{\varepsilon}} \prod_{i=1}^{N} p(\phi_i - \phi_{i-1}) \prod_{i=1}^{N-1} (\varepsilon \delta_0(d\phi_i) + d\phi_i),$$

and satisfying the Dirichlet boundary condition (1.7): $\phi_0 = aN, \phi_N = bN$ with $a, b \in \mathbb{R}^n$, where p is a probability density function on $\mathbb{R}^n$: $\int_{\mathbb{R}^n} p(x)dx = 1$. We assume that p satisfies

$$\sup_{x \in \mathbb{R}^n} e^{\lambda \cdot x} p(x) < \infty$$

for all $\lambda \in \mathbb{R}^n$. Note that μ_N^{ε} is a generalization of the Gibbs measure with a pinning defined by (1.8) when $d = 1$. In fact, we can take $p(x) = (2\pi)^{-n/2} e^{-|x|^2/2}$. More generally, we can consider the non-quadratic Hamiltonian

$$H_N(\phi) = \sum_{\langle i,j \rangle \subset D_N} V(\phi_i - \phi_j).$$

In this way, p and V are related as

$$p(x) = \frac{e^{-V(x)}dx}{\int_{\mathbb{R}} e^{-V(x)}dx}.$$

The macroscopic height function h^N is defined by (1.9) and its linear interpolation, that is,

$$h^N(x) = \frac{[Nx] - Nx + 1}{N}\phi_{[Nx]} + \frac{[Nx] - Nx}{N}\phi_{[Nx]+1}, \quad x \in D = [0, 1].$$

We define $\sigma(u)$, $u \in \mathbb{R}^n$, by the Legendre transform

$$\sigma(u) = \sup_{\lambda \in \mathbb{R}^n} \{\lambda \cdot u - \Lambda(\lambda)\}$$

of

$$\Lambda(\lambda) = \log \int_{\mathbb{R}^n} e^{\lambda \cdot x} p(x)dx.$$

Consider J defined by (1.4) with this σ, $\xi = \xi^\varepsilon$ and $D = [0, 1]$; note that ξ^ε can be defined also in this setting as in (1.10). Then, one can show the sample path LDP [32, 111, 113], which is roughly formulated as

$$\mu_N^\varepsilon(h^N \underset{L^\infty}{\sim} h) \sim e^{-NJ^*(h)}, \quad \text{as } N \longrightarrow \infty,$$

for $h \in C([0, 1], \mathbb{R}^n)$. The rate function is given by $J^*(h) = J(h) - \inf J$. Here, the notation "$h^N \underset{L^\infty}{\sim} h$" means h^N and h are close in the $L^\infty([0, 1])$-norm. When $d = 1$, the interfaces are stiff, so that spikes do not appear. Note that this result is well-known when $\varepsilon = 0$ and ϕ_N is free (instead of $\phi_N = bN$), cf., [182], and $\sigma(u) = \frac{1}{2}|u|^2$ in the Gaussian case. The concentration property implies that

$$h^N \longrightarrow \{\text{minimizers of } J\} \quad \text{as } N \longrightarrow \infty,$$

in probability in L^∞-norm.

The results on the scaling limits at criticality ($d = 1$) are summarized in the following:

Theorem 1.1 ([32, 111]). *Let $\bar{h}$ and $\hat{h}$ be two possible minimizers given in Sect. 1.1.4 and assume the condition $J(\bar{h}) = J(\hat{h})$.*

1. *When $n = 1$, the limit of h^N under μ_N^ε is $\hat{h}$: for every $\delta > 0$,*

$$\lim_{N \to \infty} \mu_N^\varepsilon(\|h^N - \hat{h}\|_\infty \leq \delta) = 1,$$

where $\|\cdot\|_\infty$ is the supremum norm on $[0, 1]$.

2. When $n = 2$, a coexistence occurs as $N \to \infty$, that is, the limit of h^N under μ_N^ε is a mixture of $\bar{h}$ and $\hat{h}$: for every $\delta > 0$ small enough,

$$\lim_{N\to\infty} \mu_N^\varepsilon(\|h^N - \hat{h}\|_\infty \leq \delta) = \hat{\lambda},$$

$$\lim_{N\to\infty} \mu_N^\varepsilon(\|h^N - \bar{h}\|_\infty \leq \delta) = \bar{\lambda},$$

with some $0 < \hat{\lambda} < 1$, $\hat{\lambda} + \bar{\lambda} = 1$.

3. When $n \geq 3$, the limit of h^N under μ_N^ε is $\bar{h}$: for every $\delta > 0$,

$$\lim_{N\to\infty} \mu_N^\varepsilon(\|h^N - \bar{h}\|_\infty \leq \delta) = 1.$$

This can be extended to the case of the free boundary condition at $\partial_R D = \{N\}$ for the microscopic system. In fact, the pinning free energy $\xi^{\varepsilon,F}$ is similar, only bit modified and the LDP is essentially the same with $\xi = \xi^{\varepsilon,F}$. Possible minimizers $\bar{h}$ and $\hat{h}$ in this setting are as in Fig. 1.9. When $n = 1$, the condition for the coexistence (criticality) $J(\bar{h}) = J(\hat{h})$ is given by $a = \pm\sqrt{\xi/2}$.

Theorem 1.2 ([32, 111]). *Assume $J(\bar{h}) = J(\hat{h})$ under the free boundary condition.*

1. *When $n = 1$, a coexistence occurs in the limit as $N \to \infty$.*
2. *When $n \geq 2$, the limit of h^N is $\bar{h}$.*

Another extension is the case with wall effect, that is, we replace $d\phi_i$ in the definition of the Gibbs measure by the Lebesgue measure $d\phi_i^+$ on $\mathbb{R}^+$. Then, the free energy is replaced by $\xi^{\varepsilon,+}$ or $\xi^{\varepsilon,F,+}$, and it is known that there exists $\varepsilon_c^+ \geq 0$ such that $\varepsilon > \varepsilon_c^+ \iff \xi^{\varepsilon,+} > 0$. Moreover,

$$d = 1, n \geq 1 \implies \varepsilon_c^+ > \varepsilon_c \ (\geq 0) \ \text{ (in particular } \varepsilon_c^+ > 0),$$

$$d = 2, n = 1 \implies \varepsilon_c^+ > 0,$$

$$d \geq 3, n = 1 \implies \varepsilon_c^+ = 0 \text{ (no transition)}.$$

Fig. 1.9 Possible minimizers (free boundary case)

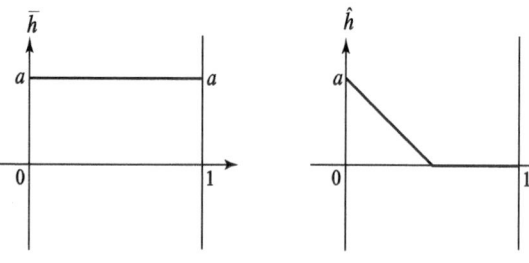

See Theorem 1.1 of [32] for $d = 1$ and Section 7.3 of [103] for $d \geq 2$, $n = 1$. We assume $a, b \in \mathbb{R}_+^n$. Then, under the balance condition $J^+(\bar{h}) = J^+(\hat{h})$, theorems analogous to Theorems 1.1 and 1.2 hold. When $d = 1$, $n \geq 1$, the critical exponents of $\xi^\varepsilon, \xi^{\varepsilon,+}$ at $\varepsilon = \varepsilon_c, \varepsilon_c^+$ can be computed, see [32, 111].

Remark 1.1. A corresponding dynamical problem is discussed by Lacoin [173] in the case where the interface takes discrete values (called SOS model) with a hard wall and an infinite pinning force to it (i.e., once the interface touches 0, it stays there forever). A conjecture is stated in Section 15 of [103] for the scaling limit of the dynamics associated with the $\nabla \varphi$-interface model in two media.

1.4 Outline of the Proof

The proofs of Theorems 1.1 and 1.2 are carried out as follows. The balance condition $J(\bar{h}) = J(\hat{h})$ implies that the (leading) exponential order of R_N vanishes, where

$$R_N = \frac{\mu_N^\varepsilon(h^N \in \text{nbd of } \hat{h})}{\mu_N^\varepsilon(h^N \in \text{nbd of } \bar{h})}, \tag{1.11}$$

and the neighborhoods are defined in the L^∞-sense; note that spikes do not appear when $d = 1$. We then compute its prefactor of the form N^α and finally obtain that

$$R_N \sim \begin{cases} N^{1-\frac{n}{2}}, & \text{in the Dirichlet boundary case,} \\ N^{\frac{1}{2}-\frac{n}{2}}, & \text{in the free boundary case,} \end{cases} \tag{1.12}$$

as $N \to \infty$. This leads to the conclusion of the theorems.

We show (1.12) in the Dirichlet boundary case. Note that, when $d \geq 3$, we will see that the ratio of two probabilities behaves as an exponential of surface order, $e^{O(N^{d-1})}$. But, when $d = 1$, this suggests that the ratio is smaller than the exponential order. Indeed, (1.12) shows that it has a polynomial (algebraic) order in N.

We denote the partition function Z_N^ε by $Z_N^{aN,bN,\varepsilon}$ to indicate the boundary conditions and denote $Z_N^{aN,bN,0}$ (i.e., $\varepsilon = 0$) simply by $Z_N^{aN,bN}$. Similarly, we denote μ_N^ε by $\mu_N^{aN,bN,\varepsilon}$ and $\mu_N^{aN,bN,0}$ simply by $\mu_N^{aN,bN}$. Then, since the condition: "$h^N \in L^\infty$-neighborhood of $\bar{h}$" implies that ϕ never touch 0 (if the line connecting a and b does not touch 0 and if the L^∞-neighborhood is chosen small enough), we see that

$$Z_N^{aN,bN,\varepsilon} \mu_N^\varepsilon(h^N \in \text{nbd of } \bar{h}) = \int 1_{\{h^N \in \text{nbd of } \bar{h}\}} \prod_{i=1}^{N} p(\phi_i - \phi_{i-1}) \prod_{i=1}^{N-1} (\varepsilon \delta_0(d\phi_i) + d\phi_i)$$

$$= \int 1_{\{h^N \in \text{nbd of } \bar{h}\}} \prod_{i=1}^{N} p(\phi_i - \phi_{i-1}) \prod_{i=1}^{N-1} d\phi_i$$

$$\sim Z_N^{aN,bN},$$

if N is large. Therefore, we have

$$R_N \sim \frac{Z_N^{aN,bN,\varepsilon}}{Z_N^{aN,bN}} \mu_N^\varepsilon (h^N \in \text{nbd of } \hat{h})$$

$$\sim \sum_{j<k} \Xi_{N,j,k}^\varepsilon \, \mu_j^{aN,0}(h^N \in \text{nbd of } \hat{h})$$

$$\times \mu_{k-j}^{0,0,\varepsilon}(h^N \in \text{nbd of } \hat{h}) \, \mu_{N-k}^{0,bN}(h^N \in \text{nbd of } \hat{h}), \tag{1.13}$$

where

$$\Xi_{N,j,k}^\varepsilon := \frac{Z_j^{aN,0} Z_{k-j}^{0,0,\varepsilon} Z_{N-k}^{0,bN}}{Z_N^{aN,bN}}$$

$$= \frac{Z_j^{aN,0} Z_{k-j}^{0,0} Z_{N-k}^{0,bN}}{Z_N^{aN,bN}} \times \frac{Z_{k-j}^{0,0,\varepsilon}}{Z_{k-j}^{0,0}} =: A \times B.$$

To derive the second line for R_N, we expand the product measure $\prod_{i=1}^{N-1}(\varepsilon\delta_0(d\phi_i) + d\phi_i)$ and show that the probability that ϕ touches 0 at most once is negligible. Thus, we may only consider the probability that ϕ touches 0 at least twice. In this event, j and k mean the first and last hitting times of ϕ at 0, respectively.

By the local central limit theorem, as $k \to \infty$ such that $k/N \to r \in (0,1]$, we have

$$Z_k^{a,b} \sim \frac{1}{(2\pi k)^{n/2}\sqrt{\det Q((b-a)/r)}} \exp\left\{-k\sigma\left(\frac{N(b-a)}{k}\right)\right\},$$

where $Q(v)$ is the covariance matrix of the Cramér transform $p_{\lambda(v)}$ of p, which has mean $v \in \mathbb{R}^n$. In fact,

$$Z_k^{a,b} = \int \prod_{i=1}^k p(\phi_i - \phi_{i-1}) \prod_{i=1}^{k-1} d\phi_i \quad (\text{with } \phi_0 = aN, \phi_k = bN)$$

$$= p^{*k}((b-a)N).$$

We apply the Cramér transform $p_\lambda(x) = e^{\lambda x - \Lambda(\lambda)} p(x)$ to replace $(b-a)N$ by 0. In this way, we obtain the asymptotic behavior of A:

$$A \sim \left(\frac{N}{j(k-j)(N-k)}\right)^{n/2} \exp\left\{-N\tilde{f}(s_1, s_2)\right\}, \tag{1.14}$$

where $s_1 = j/N$, $s_2 = 1 - k/N$ and

$$\tilde{f}(s_1, s_2) = s_1 \sigma \left(-\frac{a}{s_1} \right) + s_2 \sigma \left(\frac{b}{s_2} \right) + (1 - s_1 - s_2)\sigma(0) - \sigma(b - a).$$

On the other hand, applying the renewal theory based on the renewal equation for $Z_N^{0,0,\varepsilon}$, i.e.,

$$Z_N^{0,0,\varepsilon} = Z_N^{0,0} + \varepsilon \sum_{i=1}^{N-1} Z_i^{0,0} Z_{N-i}^{0,0,\varepsilon}, \quad N \geq 2,$$

with $Z_1^{0,0,\varepsilon} = Z_1^{0,0} = 1$, we can show that

$$B \sim (k - j)^{n/2} \exp \left\{ N\xi^\varepsilon (1 - s_1 - s_2) \right\}. \tag{1.15}$$

Indeed, by the definition of ξ^ε, we have

$$\frac{Z_\ell^{0,0,\varepsilon}}{Z_\ell^{0,0}} \sim \exp \left\{ \ell \xi^\varepsilon \right\}, \tag{1.16}$$

as $\ell \to \infty$, but (1.15) gives much more precise asymptotic behavior than (1.16).

Noting that the three probabilities in the right-hand side of (1.13) are close to 1 by the LDP, the asymptotics (1.14) and (1.15) of A and B yield

$$R_N \sim \left(\frac{N}{N^3} \right)^{n/2} \times N^{n/2} \times \sum_{j \sim t_1 N, k \sim (1 - t_2)N} e^{-Nf(s_1, s_2)}$$

$$\sim N^{-n/2} \times (\sqrt{N})^2 = N^{-n/2+1}, \tag{1.17}$$

where $f(s_1, s_2) \sim c_1(t_1 - s_1)^2 + c_2(t_2 - s_2)^2$ with some $c_1, c_2 > 0$; the top term of f vanishes because of the balance condition. Indeed,

$$\sum_{j \sim t_1 N} e^{-Nf(s_1, s_2)} \sim \sum_j e^{-c_1 N(j/N)^2}$$

$$= \sum_j e^{-c_1(j/\sqrt{N})^2} \sim \sqrt{N} \int_{\mathbb{R}} e^{-c_1 x^2} dx,$$

so that we have the factor $\sqrt{N}$ in (1.17). This is a contribution of the fluctuation around the first (or last) touching point t_1 (or t_2), which we called x_1 (or $1 - x_2$) before. Note that we have only one $\sqrt{N}$ in the right-hand side of (1.17) in the free boundary case.

1.5   Results for $d \geq 3$, $n = 1$

We now consider a random field ϕ defined on a higher-dimensional lattice cylinder D_N. We assume $d \geq 3$ for technical reasons and discuss the case of $n = 1$ only. The microscopic model is determined by the Gibbs measure μ_N^ε with a pinning on $\mathbb{R}^{D_N^o}$ introduced in (1.8), under the boundary conditions $\phi = aN$ on $\partial_L D_N$ and $\phi = bN$ on $\partial_R D_N$ with given $a, b \in \mathbb{R}$. Actually, we take $a, b > 0$. Recall that μ_N^0 (i.e., pinning $\varepsilon = 0$) is a Gaussian measure.

The corresponding macroscopic variational problem with a pinning is stated as follows: Let $D = [0, 1] \times \mathbb{T}^{d-1}$ be the continuous cylinder. We write its coordinate as $x = (x_1, \underline{x}) \in D$. Let $\xi > 0$ be given and consider

$$J(h) = \frac{1}{2} \int_D |\nabla h(x)|^2 \, dx - \xi \big| \{x \in D; h(x) = 0\} \big|,$$

for $h : D \to \mathbb{R}$ satisfying $h(0, \underline{x}) = a, h(1, \underline{x}) = b$, where as before $|\{\cdots\}|$ stands for the Lebesgue measure on D.

The microscopic model μ_N^ε and the macroscopic energy functional J are expected to be linked by the LDP:

$$\mu_N^\varepsilon(h^N \underset{L^p}{\sim} h) \sim e^{-N^d J^*(h)} \qquad \text{as } N \longrightarrow \infty, \tag{1.18}$$

where $2 \leq p < \frac{2d}{d-2}$ and the rate function should be given by $J^*(h) = J(h) - \inf J$, with $\xi = \xi^\varepsilon$ as in (1.10).

Because of the special choice of the domain D, one can see [31] that there are only two possible candidates $\bar{h}$ and $\hat{h}$ of minimizers of J determined by

$$\bar{h}(x) = \bar{h}^{(1)}(x_1), \quad \hat{h}(x) = \hat{h}^{(1)}(x_1),$$

where $\bar{h}^{(1)}$ and $\hat{h}^{(1)}$ are possible minimizers in 1-dimensional problem considered in Sect. 1.1.4; see Fig. 1.10.

The result on the scaling limits at criticality ($d \geq 3$) is formulated as follows:

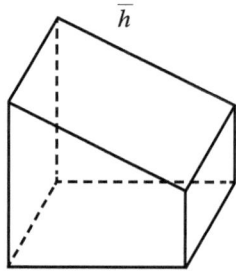
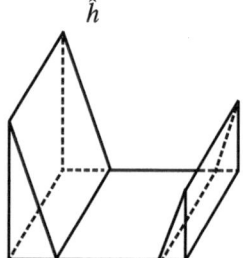

Fig. 1.10 Possible minimizers ($d \geq 2$)

Theorem 1.3 ([31]). *We assume* $J(\bar{h}) = J(\hat{h})$. *If* $d \geq 3$, $n = 1$, *and if* $\varepsilon > 0$ *is sufficiently large,* h^N *converges to* $\hat{h}$ *in* $L^1(D)$ *in probability: for every* $\delta > 0$,

$$\lim_{N \to \infty} \mu_N^\varepsilon(\|h^N - \hat{h}\|_{L^1(D)} \leq \delta) = 1.$$

In higher-dimensions, differently from the one-dimensional case, it is known that sharp spikes appear at microscopic level, as Fig. 1.7 shows. Therefore, the norm in $L^1(D)$ cannot be replaced by that of $L^\infty(D)$.

1.6 Outline of the Proof

If we consider the ratio R_N of two probabilities as in (1.11), the leading order (exponential volume order) vanishes under the balance condition (coexistence or criticality). The proof is reduced to showing the following three assertions. We consider the quantity appearing in (1.13) instead of R_N for the lower and upper bounds. The full LDP (1.18) is not yet known. Instead, we prove the concentration on $\{\hat{h}, \bar{h}\}$.

- Lower bound (surface order): For every $0 < \alpha < 1$ and $1 \leq p \leq 2$,

$$p_N := \frac{Z_N^{aN,bN,\varepsilon}}{Z_N^{aN,bN}} \mu_N^\varepsilon(\|h^N - \hat{h}\|_{L^p(D)} \leq N^{-\alpha}) \geq e^{cN^{d-1}}, \tag{1.19}$$

  with $c = c_\varepsilon (= c_{\varepsilon,\alpha}) > 0$ for $N \geq N_0$ if $\varepsilon > 0$ is large enough. The superscript aN, bN in the partition functions indicate the boundary conditions, and $Z_N^{aN,bN} = Z_N^{aN,bN,0}$ (i.e., $\varepsilon = 0$) as before.
- Upper bound (capacity order): There exists $\alpha_0 > 0$ such that

$$q_N := \frac{Z_N^{aN,bN,\varepsilon}}{Z_N^{aN,bN}} \mu_N^\varepsilon(\|h^N - \bar{h}\|_{L^p(D)} \leq (\log N)^{-\alpha_0} \leq 2, \tag{1.20}$$

  for $N \geq N_0$. (Note that if $\|\cdot\|_{L^p}$ could be replaced by $\|\cdot\|_{L^\infty}$, then "$q_N \leq 1$" is trivial. But this is not the case when $d \geq 2$ because of spikes, as we pointed out.)
- LD type estimate: There exists $\alpha_1 > 0$ such that

$$\lim_{N \to \infty} \mu_N^\varepsilon\left(\text{dist}_{L^1}(h^N, \{\hat{h}, \bar{h}\}) \geq N^{-\alpha_1}\right) = 0. \tag{1.21}$$

1.6.1 Proof of the Lower Bound

The estimate given in this part is rather rough and we need to assume that the strength of pinning $\varepsilon > 0$ is large enough.

The idea for proving (1.19) is as follows. We first restrict the event in consideration further to the event that the microscopic heights are totally pinned (i.e., $\phi = 0$)

on the microscopic bands γ_L, γ_R which correspond to the macroscopic first and last $(d-1)$-dimensional hyperplanes (points s_1^L, s_1^R if $d=1$) of $\hat{h}$ touching 0, see below. Then, we use the Markov property of the field and separate the domains. We apply the LDP to estimate the probabilities on each domain; in the middle domain, we apply the FKG type argument to remove the effect of the pinning. Thus, the problem is reduced to obtaining estimates on partition functions (on box domains) with boundary conditions. However, thanks to the Gaussian property (and especially because the domains are boxes), this is further reduced to obtaining estimates on those with null boundary conditions. Because of the criticality assumption (or the balance condition) $J(\bar{h}) = J(\hat{h})$, the leading exponential volume order term vanishes. We need to show that the next surface order term has a positive coefficient, which is the case if ε is large enough. We apply the random walk representation of the partition functions with null boundary conditions.

Let us discuss in more details. We divide D_N° as $D_N^\circ = A_L \cup \gamma_L \cup B \cup \gamma_R \cup A_R$:

$$A_L = \left([1, Ns_1^L - K - 1] \cap \mathbb{Z}\right) \times \mathbb{T}_N^{d-1},$$

$$\gamma_L = \left([Ns_1^L - K, Ns_1^L] \cap \mathbb{Z}\right) \times \mathbb{T}_N^{d-1},$$

$$B = \left([Ns_1^L + 1, Ns_1^R] \cap \mathbb{Z}\right) \times \mathbb{T}_N^{d-1},$$

$$\gamma_R = \left([Ns_1^R + 1, Ns_1^R + K] \cap \mathbb{Z}\right) \times \mathbb{T}_N^{d-1},$$

$$A_R = \left([Ns_1^R + K + 1, N - 1] \cap \mathbb{Z}\right) \times \mathbb{T}_N^{d-1},$$

with some $K \in \mathbb{N}$, where s_1^L and $s_1^R \in (0,1)$ are the first and last s's such that $\hat{h}^{(1)}(s) = 0$. Then, we can estimate

$$Z_N^{aN,bN} p_N = \int 1_{\{\|h^N - \hat{h}\|_{L^p(D)} \leq \delta\}} e^{-H_N(\phi)} \prod_{i \in D_N^\circ} \left(\varepsilon \delta_0(d\phi_i) + d\phi_i\right)$$

$$\geq \int 1_{\{\|h^N - \hat{h}\|_{L^p(D)} \leq \delta, \phi_i = 0 \text{ on } \gamma_L \cup \gamma_R, \phi_i \neq 0 \text{ on } A_L \cup A_R\}} e^{-H_N(\phi)} \prod_{i \in D_N^\circ} \left(\varepsilon \delta_0(d\phi_i) + d\phi_i\right)$$

$$= \varepsilon^{|\gamma_L| + |\gamma_R|} \int 1_{\{\|h^N - \hat{h}\|_{L^p(D_L)} \leq \delta\}} e^{-H_N(\phi)} \prod_{i \in A_L} d\phi_i$$

$$\times \int 1_{\{\|h^N - \hat{h}\|_{L^p(D_M)} \leq \delta\}} e^{-H_N(\phi)} \prod_{i \in B} \left(\varepsilon \delta_0(d\phi_i) + d\phi_i\right)$$

$$\times \int 1_{\{\|h^N - \hat{h}\|_{L^p(D_R)} \leq \delta\}} e^{-H_N(\phi)} \prod_{i \in A_R} d\phi_i$$

$$= \varepsilon^{|\gamma_L| + |\gamma_R|} Z_{A_L}^{aN,0} \mu_{A_L}^{aN,0}(\|h^N - \hat{h}\|_{L^p(D_L)} \leq \delta)$$

$$\times Z_B^{0,0,\varepsilon} \mu_B^{0,0,\varepsilon}(\|h^N - \hat{h}\|_{L^p(D_M)} \leq \delta) \times Z_{A_R}^{0,bN} \mu_{A_R}^{0,bN}(\|h^N - \hat{h}\|_{L^p(D_R)} \leq \delta),$$

where $\delta = N^{-\alpha}$, the definitions of the measures $\mu_{A_L}^{aN,0}, \mu_B^{0,0,\varepsilon}, \mu_{A_R}^{0,bN}$ are similar to those used above and clear, $Z_B^{0,0,\varepsilon}$ is the normalization constant for $\mu_B^{0,0,\varepsilon}$, and D_L, D_M, D_R are the macroscopic regions corresponding to A_L, B, A_R, respectively. In this way, restricting the probability appearing in p_N to the event that $\phi_i = 0$ for all $i \in \gamma_L \cup \gamma_R$, the Markov property of $\mu_N^{aN,bN,\varepsilon}$ shows that

$$p_N \geq \frac{Z_{A_L}^{aN,0} Z_B^{0,0,\varepsilon} Z_{A_R}^{0,bN}}{Z_N^{aN,bN}} \varepsilon^{|\gamma_L|+|\gamma_R|} \times \mu_{A_L}^{aN,0}(\|h^N - \hat{h}\|_{L^p(D_L)} \leq \delta)$$

$$\times \mu_B^{0,0,\varepsilon}(\|h^N - \hat{h}\|_{L^p(D_M)} \leq \delta) \mu_{A_R}^{0,bN}(\|h^N - \hat{h}\|_{L^p(D_R)} \leq \delta).$$

However, the LDP for $\mu_{A_L}^{aN,0}, \mu_B^{0,0}, \mu_{A_R}^{0,bN}$ and the FKG type argument applied for $\mu_B^{0,0,\varepsilon}$, which implies that

$$\mu_B^{0,0,\varepsilon}(\|h^N - \hat{h}\|_{L^p(D_M)} \leq \delta) \leq \mu_B^{0,0}(\|h^N - \hat{h}\|_{L^p(D_M)} \leq \delta),$$

show that the three probabilities in the right-hand side are all close to 1 for N sufficiently large. Thus, for every $c > 0$, we obtain

$$p_N \geq \frac{Z_{A_L}^{aN,0} Z_B^{0,0,\varepsilon} Z_{A_R}^{0,bN}}{Z_N^{aN,bN}} \varepsilon^{|\gamma_L|+|\gamma_R|} \times (1 - c) =: \Xi_N \times (1 - c),$$

for $N \geq N_0$ with some $N_0 \in \mathbb{N}$.

Next, we need to derive estimates on the partition functions. By the Gaussian property, we have

$$Z_N^{aN,bN} = \exp\left\{-\frac{N^d}{2}(a - b)^2\right\} Z_N^{0,0},$$

$$Z_{A_L}^{aN,0} = \exp\left\{-\frac{a^2 N^d}{2(s_1^L - K/N)}\right\} Z_{A_L}^{0,0},$$

$$Z_{A_R}^{0,bN} = \exp\left\{-\frac{b^2 N^d}{2(1 - s_1^R - K/N)}\right\} Z_{A_R}^{0,0}.$$

In fact, in the integration

$$Z_N^{aN,bN} = \int_{\mathbb{R}^{D_N^\circ}} e^{-H_N(\phi)} d\phi,$$

we decompose $\phi = \widetilde{\phi} + \bar{\phi}$, where ϕ satisfies (aN, bN)-boundary condition, $\widetilde{\phi}$ satisfies null boundary condition and $\bar{\phi}$ is the mean, which is affine with (aN, bN)-boundary condition. Then, we can decompose

$$H_N(\phi) = \frac{1}{2} \sum |\nabla \widetilde{\phi} + \nabla \bar{\phi}|^2$$

$$= \frac{1}{2} \sum |\nabla \widetilde{\phi}|^2 + \sum \nabla \widetilde{\phi} \cdot \nabla \bar{\phi} + \frac{1}{2} \sum |\nabla \bar{\phi}|^2,$$

where $\nabla \phi \equiv \nabla \phi(b) := \phi_i - \phi_j$ for $b = \langle i, j \rangle$. However,

$$\sum \nabla \widetilde{\phi} \cdot \nabla \bar{\phi} = (a-b)(\widetilde{\phi}|_{\partial_L D} - \widetilde{\phi}|_{\partial_R D}) = 0$$

and, since $\nabla \bar{\phi}(b) = a - b$ for i_1-directed b and $= 0$ for b with other directions,

$$\frac{1}{2} \sum |\nabla \bar{\phi}|^2 = \frac{N^d}{2}(a-b)^2.$$

Thus, we obtain the first equality. The others are shown similarly.

However, we have the expansion $1/(s_1^L - K/N) = 1/s_1^L + KN^{-1}/(s_1^L)^2 + O(N^{-2})$ and a similar one for $1/(1 - s_1^R - K/N)$ as $N \to \infty$. Consequently,

$$\Xi_N = \exp \{f(a,b)N^d - K\tilde{f}(a,b)N^{d-1} - O(N^{d-2})\} \Xi_N^0,$$

where Ξ_N^0 is Ξ_N with $a = b = 0$, and

$$f(a,b) = \frac{1}{2}(a-b)^2 - \frac{a^2}{2s_1^L} - \frac{b^2}{2(1-s_1^R)},$$

$$\tilde{f}(a,b) = \frac{a^2}{2(s_1^L)^2} + \frac{b^2}{2(1-s_1^R)^2}.$$

Note that $f(a,b) = J(\bar{h}) - J(\hat{h}) - \xi^\varepsilon(s_1^R - s_1^L)$, and $\tilde{f}(a,b) = 2\xi^\varepsilon$ holds from Young's relation for the angles of $\hat{h}$ at $s = s_1^L$ and s_1^R: $a/s_1^L = b/(1-s_1^R) = \sqrt{2\xi^\varepsilon}$.

To estimate Ξ_N^0 from below, we use the random walk representation of partition functions, which yields

$$\frac{Z_{A_L}^{0,0} Z_{A_R}^{0,0}}{Z_N^{0,0}} \geq \exp \{\hat{q}^0(|A_L| + |A_R| - |D_N^\circ|) - C_1 N^{d-1}\}, \tag{1.22}$$

where

$$\hat{q}^0 = \frac{1}{2}\left(\log \frac{\pi}{d} + q\right), \quad q = \sum_{k=1}^{\infty} \frac{1}{2k} P_0^{RW^d}(\eta_{2k} = 0).$$

In fact, when N is even, the random walk representation of the partition functions with null boundary condition is given by

$$\log Z_A^0 = \frac{1}{2}\left(|A| \log \frac{\pi}{d} + I\right), \tag{1.23}$$

where

$$I = \sum_{k \in A} \sum_{n=1}^{\infty} \frac{1}{2n} P_k(\eta_{2n} = k, \tau_A > 2n),$$

(η_n) is a random walk on D_N, and τ_A is the first exit time of η from A; note that, since N is even, $P_k(\eta_{2n-1} = k) = 0$. Relation (1.23) can be shown as follows:

$$Z_A^0 = (2\pi)^{\frac{|A|}{2}} \left(\det(-\Delta_A)\right)^{-\frac{1}{2}},$$

where $\Delta_A = 2d(P_A - I)$ and

$$(P_A)_{ij} = \begin{cases} 1, & |i-j| = 1, \ i,j \in A, \\ 0, & \text{otherwise.} \end{cases}$$

Then, denoting the eigenvalues of P_A by $\{\mu_k\}_{k=1}^{|A|}$ and applying the Taylor expansion of $\log(1-x)$, we have

$$\begin{aligned}
\log Z_A^0 &= \frac{|A|}{2} \log 2\pi - \frac{1}{2} \sum_{k=1}^{|A|} \log\left(2d(1-\mu_k)\right) \\
&= \frac{|A|}{2} \log \frac{\pi}{d} - \frac{1}{2} \sum_{k=1}^{|A|} \log(1-\mu_k) \\
&= \frac{|A|}{2} \log \frac{\pi}{d} + \frac{1}{2} \sum_{n=1}^{\infty} \frac{1}{n} \sum_{k=1}^{|A|} (\mu_k)^n \\
&= \frac{|A|}{2} \log \frac{\pi}{d} + \sum_{n=1}^{\infty} \frac{1}{2n} \operatorname{Tr} P_A^n,
\end{aligned}$$

and this implies (1.23). From (1.23), one can obtain (1.22), see Section 4, Lemma 2.3 and Remark 2.8-(2) in [31].

Moreover, by the decoupling estimate [33], we have

$$Z_B^{0,0,\varepsilon} \geq \exp\left\{\hat{q}^{\varepsilon}|B| - \frac{3}{2}cN^{d-1}\right\},$$

where $\hat{q}^{\varepsilon} = \hat{q}^0 + \xi^{\varepsilon}$, $c = G(0,0)$, and G is the Green function of the d-dimensional random walk on $\mathbb{Z}^d$; recall $d \geq 3$.

Summarizing these estimates, the leading term of $e^{O(N^d)}$ is given by $f(a, b) + \xi^\varepsilon(s_1^R - s_1^L)$ (by noting that the coefficient of $\hat{q}^0$ vanishes and $\xi^\varepsilon = \hat{q}^\varepsilon - \hat{q}^0$), and this vanishes by the balance condition $J(\bar{h}) = J(\hat{h})$. Then, by noting that $\xi^\varepsilon \leq \log 2\varepsilon$ (for $\varepsilon \geq 1$), we have

$$\log p_N \geq \left((2K + 1)(\log \varepsilon - \hat{q}^0) - 2K\xi^\varepsilon - C_1 - \frac{3}{2}c\right)N^{d-1} - O(N^{d-2})$$

$$\geq \left(\log \varepsilon - (2K + 1)\hat{q}^0 - 2K\log 2 - C_1 - \frac{3}{2}c\right)N^{d-1} - O(N^{d-2}).$$

Since the coefficient of N^{d-1} is positive if $\varepsilon > 0$ is large enough, this completes the proof of the lower bound.

1.6.2   Proof of the Upper Bound

The proof of the upper bound (1.20) is outlined as follows. We first expand the product measure of $\varepsilon\delta_0(d\phi_i) + d\phi_i$ in i and denote i's for $d\phi_i$ by A and i's for $\delta_0(d\phi_i)$ by $A^c := D_N^\circ \setminus A$ so that $\phi_i = 0$ on A^c. We call A the wet region and A^c the dry region, since here $\phi_i = 0$ means that i is not covered by water. Then, because we are going to estimate the probability of the event that h^N is close to $\bar{h}$ (which is positive), the dry region A^c cannot be too large. Indeed, we observe that we may consider the case $|A^c| \leq (N/\log N)^d$ only. Then, ignoring the probability (by estimating it by 1 from above), the problem is reduced to estimating the sum of the ratio of two partition functions, Z_A/Z_N; see the quantity r_N described below. However, by the Gaussian property, the ratio of the partition functions (which can be regarded as the marginal density evaluated at 0 on A^c, see (1.24)) can be explicitly computed, and we obtain their estimates using the capacity, see (1.25). Based on the isoperimetric inequality and a result of Kaimanovich [158], the capacity can be estimated from below by $(d-2)/d$-th power of the volume of the set, see (1.26). From this, we can show that $r_N = 1 + o(1)$ as $N \to \infty$, which is bounded by 2 for large N. Note that the upper bound is almost 1. If we can take the L^∞-norm in place of the L^p-norm in (1.20) (which is actually not possible because spikes appear if $d \geq 2$), the upper bound "$q_N \leq 1$" is trivial. In fact, on the event $E = \{\|h^N - \bar{h}\|_{L^\infty(D)} \leq (\log N)^{-\alpha_0}\}$, we always have $\phi_i \neq 0$ for all i, so that

$$Z_N^{aN,bN,\varepsilon}\mu_N^\varepsilon(E) = \int_E e^{-H_N(\phi)} \prod d\phi_i \leq \int e^{-H_N(\phi)} \prod d\phi_i = Z_N^{aN,bN},$$

which implies "$q_N \leq 1$". However, if we use the L^p-norm, we don't have "$\phi_i \neq 0$ for all i".

We expand the measure $\prod_{i\in D_N^\circ}\left(\varepsilon\delta_0(d\phi_i)+d\phi_i\right)$ as $\sum_{A\subset D_N^\circ}\varepsilon^{|A^c|}\prod_{i\in A}d\phi_i$ and get

$$q_N = \sum_{A\subset D_N^\circ}\varepsilon^{|A^c|}\frac{Z_A^{aN,bN,0}}{Z_N^{aN,bN}}\mu_A^{aN,bN,0}(\|h^N-\bar{h}\|_{L^p(D)}\leq\delta),$$

with $\delta=(\log N)^{-\alpha_0},\alpha_0>d/p$. Here, $\mu_A^{aN,bN,0}$ denotes the Gibbs measure without pinning and with the boundary conditions 0 on A^c (i.e., $\phi_i=0$ on A^c) and (1.7) on ∂D_N. If $|A^c|\geq(N/\log N)^d$, then $h^N=0$ on $\frac{1}{N}A^c$, so that $\|h^N-\bar{h}\|_{L^p}\geq c(\log N)^{-d/p}$ and $\mu_A^{aN,bN,0}(\|h^N-\bar{h}\|_{L^p(D)}\leq\delta)=0$. Therefore, we can show that

$$r_N := \sum_{A\subset D_N^\circ:|A^c|\leq(N/\log N)^d}\varepsilon^{|A^c|}\frac{Z_A^{aN,bN,0}}{Z_N^{aN,bN}}\leq 2.$$

To this end, we first introduce several notations:

$$P_{A^c}(i,j)=P_i(\text{RW enters }A^c\text{ at }j\text{ before reaching }\partial D_N),$$

$$p_L(i)=P_i(\text{RW does not return to }A^c\text{ and leaves }D_N^\circ\text{ via }\partial_L D_N),$$

$$p_R(i)=P_i(\text{RW does not return to }A^c\text{ and leaves }D_N^\circ\text{ via }\partial_R D_N),$$

$$e_{A^c}(i)=1-\sum_{j\in A^c}P_{A^c}(i,j)\big(=p_L(i)+p_R(i)\big),$$

$$\text{cap}_{D_N}(A^c)=\sum_{i\in A^c}e_{A^c}(i),$$

where RW denotes the random walk on $\mathbb{Z}\times\mathbb{T}_N^{d-1}$ starting at i. The numbers $e_{A^c}(i)$ and $\text{cap}_{D_N}(A^c)$ are called the escape probability and the capacity. Then, by the Gaussian property, we get

$$\frac{Z_A^{aN,bN,0}}{Z_N^{aN,bN}}=\frac{1}{Z_N^{aN,bN}}\int_{\mathbb{R}^{D_N^\circ}}e^{-H_N(\phi)}\prod_{i\in A}d\phi_i\prod_{i\in A^c}\delta_0(d\phi_i)$$

$$=\frac{1}{\sqrt{(2\pi)^{|A^c|}\det\Gamma_{A^c}}}\exp\left\{-d\langle m,(I-P_{A^c})m\rangle_{A^c}\right\}, \tag{1.24}$$

which is a marginal density of $\frac{1}{Z_N^{aN,bN}}e^{-H_N(\phi)}$ on A^c (i.e., integrated over A) evaluated at $\phi_i=0$ on A^c, where $P_{A^c}=(P_{A^c}(i,j))_{i,j\in A^c}$, $\Gamma_{A^c}=(I-P_{A^c})^{-1}$, and m is the linear interpolant on D_N° satisfying the boundary condition (1.7). The denominator in the right-hand side of (1.24) can be estimated by $e^{-d|A^c|}$ from below.

On the other hand, for the numerator we see that

$$m(i) = \sum_{j \in A^c} P_{A^c}(i,j)\, m(j) + p_L(i)\, aN + p_R(i)\, bN.$$

Therefore, we obtain

$$(I - P_{A^c})m(i) \geq \min(a,b)\, N e_{A^c}(i), \quad i \in A^c.$$

Since we also have $m(i) \geq \min(a,b)\,N$, this combined with the estimate on the denominator leads to

$$\frac{Z_A^{aN,bN,0}}{Z_N^{aN,bN}} \leq e^{d|A^c| - cN^2 \mathrm{cap}_{D_N}(A^c)}, \tag{1.25}$$

with some $c > 0$.

However, we have the capacity bound:

$$\mathrm{cap}_{D_N}(A^c) \geq c|A^c|^{(d-2)/d}. \tag{1.26}$$

In fact, this can be shown as follows. First we divide the cylinder D_N into several boxes and show that (1.26) follows once we can show this for A which is not connected in the vertical direction (i.e., the direction of $\mathbb{T}_N^{d-1}$). This reduces the problem on $\mathbb{Z}^d$. But (1.26) on $\mathbb{Z}^d$ in place of D_N follows from the isoperimetric inequality

$$|\partial A| \geq c|A|^{(d-1)/d},$$

combined with a result from [158].

Now, setting $\chi(k) = \sharp\{A; |A^c| = k\}$, we have from (1.25) and (1.26) that

$$r_N \leq \sum_{k=0}^{(N/\log N)^d} \chi(k)\varepsilon^k e^{dk - \bar{c}N^2 k^{(d-2)/d}}.$$

The term with $k = 0$ is 1, while the sum of other terms ≤ 1 if N is large, since

$$\chi(k) = \binom{N^d}{k} = \frac{N^d(N^d - 1)\cdots(N^d - k + 1)}{k!} \leq (N^d)^k = e^{dk \log N}.$$

This concludes the proof of the upper bound.

1.6.3 Proof of the Large Deviation Type Estimate

We first outline the proof presented in this part. We introduce *mesoscopic regions*, which are small in macroscopic size, but large in microscopic size. The reason for introducing such sets is twofold. First, they are large enough for the averaging effect to yield ξ^ε. Second, the number of such sets $\#\{\widehat{\mathscr{B}}_N\}$ is not large, so that we can estimate the sum of probabilities by the maximum over $B \in \widehat{\mathscr{B}}_N$ as in (1.30). Then, in (1.30), we consider two cases for B separately: If B has a small mesoscopic energy $E_N^*(B)$, then by the mesoscopic stability result stated in (1.29), the event $\mathrm{dist}_{L^1}(h^N, \{\bar{h}, \hat{h}\}) \geq N^{-\alpha}$ cannot happen or happens rarely, so that the probability is small. On the other hand, if $E_N^*(B)$ is large, since the probability behaves as $e^{-E_N^*(B)}$, this is again small.

We first show the stability of J at macroscopic level:

$$J^*(h) \leq \delta \implies \mathrm{dist}_{L^1}(h, \{\bar{h}, \hat{h}\}) \leq C\sqrt{\delta}. \tag{1.27}$$

Next, we introduce mesoscopic regions as follows: Given $0 < \beta < 1$, we divide D_N into $N^{d(1-\beta)}$ subboxes of sidelength N^β. We write $\mathscr{B}_N$ for the set of these subboxes, and $\widehat{\mathscr{B}}_N$ for the set of unions of boxes in $\mathscr{B}_N$. For $B \in \widehat{\mathscr{B}}_N$, which is called a mesoscopic region, set

$$E_{N,0}(B) = \inf_{\phi \in \mathbb{R}^{D_N}:(1.28)} H_N(\phi),$$

$$E_N(B) = E_{N,0}(B) - \xi^\varepsilon |B^c|,$$

$$E_N^*(B) = E_N(B) - \min_{B \in \widehat{\mathscr{B}}_N} E_N(B),$$

where $H_N(\phi)$ is defined in (1.6) and the infimum is taken over all $\phi \in \mathbb{R}^{D_N}$ satisfying the condition:

$$\phi_i = \begin{cases} aN, & \text{if } i \in \partial_L D_N, \\ bN, & \text{if } i \in \partial_R D_N, \\ 0, & \text{if } i \in D_N^\circ \setminus B. \end{cases} \tag{1.28}$$

Then, the stability (1.27) of J at macroscopic level can be extended to the mesoscopic level: For $\alpha > 0$, there exists $\delta = \delta(\alpha) > 0$ such that, if N is large enough,

$$E_N^*(B) \leq N^{d-\delta} \implies \mathrm{dist}_{L^1}(h_B^N, \{\bar{h}, \hat{h}\}) \leq N^{-\alpha}, \tag{1.29}$$

holds for $B \in \widehat{\mathscr{B}}_N$, where h_B^N is defined as follows: Let $\bar{\phi}^B$ be the harmonic function on B subject to the condition (1.28). Then, the macroscopic profile h_B^N is defined from the microscopic profile $\bar{\phi}^B$ by polilinearly interpolating $\frac{1}{N}\bar{\phi}_{[Nx]}^B$, $x \in D$.

We further introduce mesoscopic wetted region: Fix $\gamma > 0$ and

$$\mathcal{M}_N \equiv \mathcal{M}_N(\phi) \overset{\text{def}}{=} \bigcup \left\{ C \in \mathcal{B}_N; \, \phi_C^{\text{cg},\beta,N} \geq N^\gamma \right\},$$

where, for $C \in \mathcal{B}_N$,

$$\phi_C^{\text{cg},\beta,N} \overset{\text{def}}{=} N^{-d\beta} \sum_{j \in C} \phi_j.$$

We now decompose and estimate the probability under consideration as follows:

$$\mu_N^\varepsilon \left(\text{dist}_{L^1}(h^N, \{\bar{h}, \hat{h}\}) \geq N^{-\alpha} \right)$$

$$= \sum_{B \in \widehat{\mathcal{B}}_N} \mu_N^\varepsilon \left(\text{dist}_{L^1}(h^N, \{\bar{h}, \hat{h}\}) \geq N^{-\alpha}, \, \mathcal{M}_N = B \right)$$

$$\leq \sharp\{\widehat{\mathcal{B}}_N\} \max_{B \in \widehat{\mathcal{B}}_N} \mu_N^\varepsilon \left(\text{dist}_{L^1}(h^N, \{\bar{h}, \hat{h}\}) \geq N^{-\alpha}, \, \mathcal{M}_N = B \right). \qquad (1.30)$$

Since the number of mesoscopic regions $\sharp\{\widehat{\mathcal{B}}_N\} = e^{N^{d(1-\beta)} \log 2}$ is sub-exponential in N^d, if one can show that the above probability for each B is bounded by $e^{-N^{d-\delta}}$ with some $\delta < d\beta$, we obtain the conclusion.

If $E_N^*(B) \leq N^{d-\delta'}$, by mesoscopic stability, this event is hard to happen, so that the probability can be bounded by $e^{-N^{d-\delta}}$ with a suitable choice of δ'. On the other hand, if $E_N^*(B) \geq N^{d-\delta'}$, this probability is bounded by

$$\mu_N^\varepsilon \left(\mathcal{M}_N = B \right) \asymp e^{-E_N^*(B)} \leq e^{-N^{d-\delta'}}$$

by the mesoscopic averaging effect, under which the free energy ξ^ε arises.

The last part of the above explanation is rough. In the course of the proof, we use a super-exponential estimate, an analysis of super-harmonic functions on D_N, and the so-called volume filling lemma. To avoid technical difficulties caused by the (aN, bN)-boundary condition, we actually consider an extended region and replace it by the null boundary condition.

Remark 1.2. Several questions remain unsolved in the higher-dimensional setting. The case $d = 2$ is unsolved because the Green function diverges. But we believe the limit should be $\hat{h}$, since this is the case for both $d = 1$ and $d \geq 3$. When $d \geq 3$, we assume that $\varepsilon > 0$ is sufficiently large, but this assumption could be removed. We need to refine the proof of the lower bound.

Other questions: What happens for a general domain D or D_N? What happens when $d \geq 3$ and $n \geq 2$? How about the non-Gaussian case? What happens for different boundary conditions, as we discussed when $d = 1$? What happens when the set of minimizers is continuous (as the example on p.1010 of [111])? How about the corresponding dynamical problems? See Remark 1.1.

Chapter 2
Dynamic Young Diagrams

Young diagrams can be regarded as decreasing interfaces which separate two distinct phases, see Figs. 2.1 and 2.17 below. Such random interfaces appear also in zero-temperature Ising models at the corner, see Fig. 2.3. The goal of this chapter is to study the dynamics of random Young diagrams, sometimes called SOS (solid on solid) dynamics, mostly in two-dimensional (2D) case, which is naturally associated with the grand canonical and canonical ensembles both in uniform and restricted uniform statistics, introduced by Vershik [213]. We first recall ensembles of two-dimensional Young diagrams and their scaling limits, that is, the law of large numbers, which leads to the so-called Vershik curves in the limits, their fluctuations and large deviation principle. Then, we introduce the corresponding dynamics, and establish its space-time scaling limits such as hydrodynamic limit (law of large numbers, obtained with Sasada [114]) and non-equilibrium fluctuations (obtained with Sasada, Sauer, and Xie [115]) for non-conservative case, i.e., for the dynamics associated with the grand canonical ensembles. Vershik curves can be recovered in these dynamic scaling limits. We also discuss the dynamics associated with the canonical ensembles, which has a conservation law. We finally discuss the three-dimensional (3D) case. Cerf and Kenyon [45] derived the limit surface called Wulff shape, which is characterized by a certain variational formula under uniform statistics. We discuss the corresponding dynamic problem.

2.1 Static Theory of 2D Young Diagrams

We first introduce four types of randomness for the class of 2D Young diagrams called uniform/restricted uniform (or Bose/Fermi) and canonical/grand canonical ensembles described by probability measures $\mu_U^n, \mu_U^\varepsilon, \mu_R^n, \mu_R^\varepsilon$, respectively. Then, we discuss scaling limits under these ensembles as the size of diagrams grows.

© The Author(s) 2016   29
T. Funaki, *Lectures on Random Interfaces*, SpringerBriefs in Probability
and Mathematical Statistics, DOI 10.1007/978-981-10-0849-8_2

The limit shapes of diagrams appearing in the law of large numbers (LLN) are called Vershik curves. Fluctuations and the large deviation principle (LDP) are also discussed.

2.1.1 Ensembles of 2D Young Diagrams

For each $n \in \mathbb{N}$, let $\mathscr{P}_n$ be the set of all partitions of n or sequences $p = (p_i)_{i \in \mathbb{N}}$ satisfying $p_1 \geq p_2 \geq \cdots \geq p_i \geq \cdots$, $p_i \in \mathbb{Z}_+$ and $n(p) := \sum_{i \in \mathbb{N}} p_i = n$. We use the notation: $\mathbb{Z}_+ = \{0, 1, 2, \ldots\}$, $\mathbb{N} = \{1, 2, 3, \ldots\}$, $\mathbb{R}_+ = [0, \infty)$, and $\mathbb{R}_+^\circ = (0, \infty)$. Let $\mathscr{Q}_n$ be the set of all sequences $q = (q_i)_{i \in \mathbb{N}} \in \mathscr{P}_n$ satisfying $q_i > q_{i+1}$ if $q_i > 0$. For $n = 0$, we define $\mathscr{P}_0 = \mathscr{Q}_0 = \{0\}$, where 0 is the sequence with $p_i = 0$ for all $i \in \mathbb{N}$. The unions of $\mathscr{P}_n$ and $\mathscr{Q}_n$ in $n \in \mathbb{Z}_+$ are denoted by $\mathscr{P}$ and $\mathscr{Q}$, respectively. The height function of the Young diagram corresponding to $p \in \mathscr{P}$ is defined by

$$\psi_p(u) = \sum_{i \in \mathbb{N}} 1_{\{u < p_i\}}, \quad u \in \mathbb{R}_+. \tag{2.1}$$

In this way, we identify $p \in \mathscr{P}$ with the height function ψ_p or the Young diagram shown in Fig. 2.1 itself. In particular, the area of the Young diagram is always given by $\int_0^\infty \psi_p(u)du = n(p)$. Note that ψ_q is defined for $q \in \mathscr{Q}$, since $\mathscr{Q} \subset \mathscr{P}$.

Now, following Vershik [212], let us introduce four types of randomness, called statistics or ensembles, on the sets $\mathscr{P}_n, \mathscr{P}, \mathscr{Q}_n, \mathscr{Q}$, respectively. In fact, Vershik discussed several other ensembles, but here we concentrate on these simplest and natural ones. The first one is the *uniform (U) statistics* on $\mathscr{P}_n$ or $\mathscr{P}$, which Vershik calls Bose statistics. For $n \in \mathbb{Z}_+$, let μ_U^n be the uniform probability measure on $\mathscr{P}_n$. Moreover, for $0 < \varepsilon < 1$, let μ_U^ε be the probability measure on $\mathscr{P}$ determined by

$$\mu_U^\varepsilon(p) = \frac{1}{Z_U(\varepsilon)} \varepsilon^{n(p)}, \quad p \in \mathscr{P}, \tag{2.2}$$

where $Z_U(\varepsilon) = \prod_{k=1}^\infty (1 - \varepsilon^k)^{-1} \left(= \sum_{n=0}^\infty p(n) \varepsilon^n, p(n) = \sharp \mathscr{P}_n \right)$ is the normalization constant. The measure μ_U^ε has the property $\mu_U^\varepsilon|_{\mathscr{P}_n}(p) = \mu_U^n(p)$, $p \in \mathscr{P}$, where $\mu_U^\varepsilon|_{\mathscr{P}_n}$ stands for the conditional probability of μ_U^ε on $\mathscr{P}_n$, so that μ_U^ε is a

Uniform (Bose) case: Restricted Uniform (Fermi) case:
height $\psi \colon \mathbb{R}_+ \to \mathbb{Z}_+$

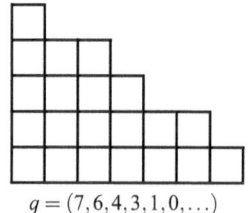

$p = (7, 6, 4, 4, 1, 0, \ldots)$ $q = (7, 6, 4, 3, 1, 0, \ldots)$

Fig. 2.1 2D young diagrams

Fig. 2.2 Height differences

Uniform (Bose) case: height difference $\eta : \mathbb{N} \to \mathbb{Z}_+$

Restricted Uniform (Fermi) case:

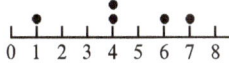

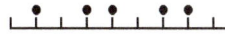

Fig. 2.3 Zero-temperature stochastic Ising model

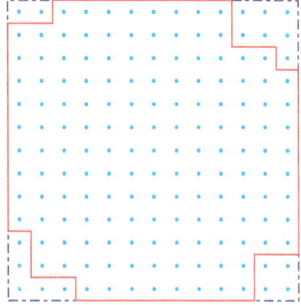

superposition of $\{\mu_U^n\}_{n \in \mathbb{Z}_+}$. The measures μ_U^n and μ_U^ε play roles similar to the canonical and grand canonical ensembles in statistical physics, respectively.

Next we introduce the *restricted uniform (RU) statistics* on $\mathcal{Q}_n$ or $\mathcal{Q}$, also called Fermi statistics. For $n \in \mathbb{Z}_+$, let μ_R^n be the uniform probability measure on $\mathcal{Q}_n$. Moreover, for $0 < \varepsilon < 1$, let μ_R^ε be the probability measure on $\mathcal{Q}$ defined by

$$\mu_R^\varepsilon(q) = \frac{1}{Z_R(\varepsilon)} \varepsilon^{n(q)}, \quad q \in \mathcal{Q}, \tag{2.3}$$

where $Z_R(\varepsilon) = \prod_{k=1}^\infty (1 + \varepsilon^k) \big(= \sum_{n=0}^\infty p_{\neq}(n)\varepsilon^n, p_{\neq}(n) = \sharp \mathcal{Q}_n \big)$ is the normalization constant, see [212]. The conditional measure $\mu_R^\varepsilon|_{\mathcal{Q}_n}$ of μ_R^ε on $\mathcal{Q}_n$ coincides with μ_R^n. The measures μ_R^n and μ_R^ε are the canonical and grand canonical ensembles in the RU (restricted uniform)-statistics, respectively. We call this the RU-case (or Fermi) because the values of corresponding height differences defined below are restricted to be 0 or 1 as we easily see in this case.

One can identify the height function $\psi : \mathbb{R}_+ \to \mathbb{Z}_+$ with its *height difference* $\eta : \mathbb{N} \to \mathbb{Z}_+$ defined by $\eta_k = \psi(k-1) - \psi(k)$. The η's corresponding to ψ's in Fig. 2.1 are given as in Fig. 2.2. Note that $\eta_k \in \mathbb{Z}_+$ in the U-case, while $\eta_k \in \{0, 1\}$ in the RU-case.

Young diagrams appear in the zero-temperature Ising model at the corner of boxes shown in the following figure, see Caputo, Martinelli, Simenhaus, and Toninelli [42], see Fig. 2.3.

2.1.2   Scaling Limits, Law of Large Numbers, and Vershik Curves

The height function of a Young diagram describes a decreasing interface at a microscopic level. The corresponding macroscopic height function is defined by

Fig. 2.4 Scaling in 2D

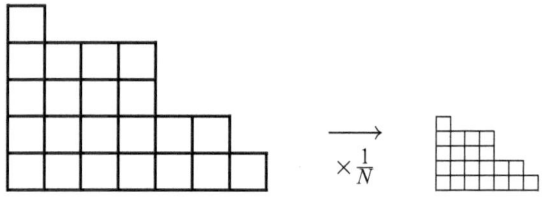

scaling as follows. For $\psi \in \mathscr{P}$ of large size with area N^2 (or with averaged area in a random case), we introduce a scaled height function $\tilde{\psi}^N$:

$$\tilde{\psi}^N(u) := \frac{1}{N} \psi(Nu), \quad u > 0, \tag{2.4}$$

by reducing the lengths both in vertical and horizontal directions by $1/N$, so that the area (or averaged area) of the scaled Young diagram determined by $\tilde{\psi}^N$ is normalized to 1, see Fig. 2.4.

For $N > 0$ such that $N^2 \in \mathbb{N}$, we consider $\psi \in \mathscr{P}$ distributed under $\mu_U^{N^2}$, i.e., the uniform canonical ensemble with area $n = N^2$, or under $\mu_R^{N^2}$, i.e., the restricted uniform canonical ensemble with area $n = N^2$. To consider under the grand canonical ensembles, for $N > 0$, choose $\varepsilon \equiv \varepsilon(N) = \varepsilon_U(N)$ or $\varepsilon_R(N) > 0$ such that

$$E^{\mu_U^{\varepsilon}}[n(\psi)] = N^2, \quad E^{\mu_R^{\varepsilon}}[n(\psi)] = N^2. \tag{2.5}$$

Namely, the averaged areas of Young diagrams under $\mu_U^{\varepsilon(N)}$ and $\mu_R^{\varepsilon(N)}$ are both N^2. The following asymptotic behaviors of $\varepsilon_U(N)$ and $\varepsilon_R(N)$ are known:

$$\varepsilon_U(N) = 1 - \frac{\alpha}{N} + O\left(\frac{\log N}{N^2}\right), \quad \alpha = \frac{\pi}{\sqrt{6}},$$

$$\varepsilon_R(N) = 1 - \frac{\beta}{N} + O\left(\frac{\log N}{N^2}\right), \quad \beta = \frac{\pi}{\sqrt{12}},$$

as $N \to \infty$, cf. [114] Lemmas 3.1 and 3.2. The first one is related to Hardy-Ramanujan formula for the Euler function $p(n) = \sharp \mathscr{P}_n$:

$$p(n) \sim \frac{1}{4\sqrt{3}n} e^{2\alpha \sqrt{n}} \quad \text{as } n \longrightarrow \infty.$$

Then, the law of large numbers (LLN) holds for $\tilde{\psi}^N$ under the four ensembles $\mu_U^{N^2}, \mu_U^{\varepsilon(N)}, \mu_R^{N^2}$ and $\mu_R^{\varepsilon(N)}$ as $N \to \infty$, see Vershik [212]:

Theorem 2.1. *The scaled height function $\tilde{\psi}^N(u)$ converges as $N \to \infty$ to $\psi_U(u)$ in probability under $\mu_U^{N^2}$ and $\mu_U^{\varepsilon(N)}$, and to $\psi_R(u)$ in probability under $\mu_R^{N^2}$ and $\mu_R^{\varepsilon(N)}$, respectively, where*

Fig. 2.5 Vershik curves

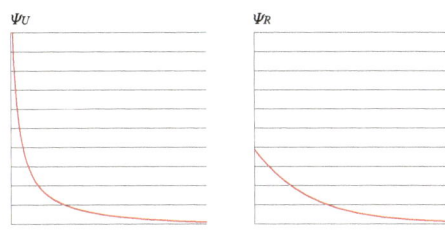

$$\psi_{\mathrm{U}}(u) = -\frac{1}{\alpha} \log\left(1 - e^{-\alpha u}\right), \quad u > 0,$$

$$\psi_{\mathrm{R}}(u) = \frac{1}{\beta} \log\left(1 + e^{-\beta u}\right), \quad u \geq 0.$$

More precisely,

$$\lim_{N \to \infty} \mu_{\mathrm{U}}^{N^2}\left(d(\tilde{\psi}^N, \psi_{\mathrm{U}}) > \delta\right) = 0,$$

and similar statements for $\mu_{\mathrm{U}}^{\varepsilon(N)}, \mu_{\mathrm{R}}^{N^2}, \mu_{\mathrm{R}}^{\varepsilon(N)}$ hold for every $\delta > 0$, where d is a distance in the space $D = \{\text{right-continuous functions on } \mathbb{R}_+^\circ\}$, equipped with the topology determined by uniform convergence on compact subsets of $\mathbb{R}_+^\circ$.

The limit shapes are called *Vershik curves*, which are the same under the canonical and grand canonical ensembles, but different in the U or RU cases, see Fig. 2.5. Note that the curve $y = \psi_{\mathrm{U}}(u)$ is implicitly expressed as $e^{-\alpha u} + e^{-\alpha y} = 1$, and $y = \psi_{\mathrm{R}}(u)$ as $e^{\beta y} - e^{-\beta u} = 1$. Vershik [212] showed the LLN under more general statistics.

2.1.3 Central Limit Theorem

Let us consider fluctuations of $\tilde{\psi}^N$ around the limits of the LLN:

$$\Psi_{\mathrm{U}}^N(u) := \sqrt{N}\left(\tilde{\psi}^N(u) - \psi_{\mathrm{U}}(u)\right), \quad u > 0,$$

$$\Psi_{\mathrm{R}}^N(u) := \sqrt{N}\left(\tilde{\psi}^N(u) - \psi_{\mathrm{R}}(u)\right), \quad u \geq 0.$$

Then one can show the following central limit theorem (CLT) under the grand canonical ensembles, cf. [115] Proposition 5.1.

Theorem 2.2. *The fluctuation fields $\Psi_{\mathrm{U}}^N(u)$ and $\Psi_{\mathrm{R}}^N(u)$ weakly converge as $N \to \infty$ to $\Psi_{\mathrm{U}}(u)$ and $\Psi_{\mathrm{R}}(u)$ under $\mu_{\mathrm{U}}^{\varepsilon(N)}$ and $\mu_{\mathrm{R}}^{\varepsilon(N)}$, respectively, where Ψ_{U} and Ψ_{R} are centered (mean 0) Gaussian processes with covariance structures*

$$C_U(u, v) = \frac{1}{\alpha} \rho_U(u \vee v), \quad u, v > 0,$$

$$C_R(u, v) = \frac{1}{\beta} \rho_R(u \vee v), \quad u, v \geq 0,$$

and $\rho_U = -\psi'_U \big(= 1/(e^{\alpha u} - 1)\big)$, $\rho_R = -\psi'_R \big(= 1/(e^{\beta u} + 1)\big)$ are the slopes of the Vershik curves, respectively, with $u \vee v = \max\{u, v\}$.

The CLT under the canonical ensembles can be deduced from that under the grand canonical ensembles by removing the effect of fluctuations of area. Here, we only cite the result of Yakubovich [221] on the weak convergence of every finite-dimensional joint distribution of the fluctuation field under the RU-canonical ensemble $\mu_R^{N^2}$.

Theorem 2.3. *Consider the fluctuation field*

$$\tilde{\Psi}_R^N(u) = \beta \sqrt{N} \big(\tilde{\psi}^N(u/\beta) - \psi_R(u/\beta)\big).$$

Then, under $\mu_R^{N^2}$, for every $0 \leq u_1 < u_2 < \cdots < u_k$, $\{\tilde{\Psi}_R^N(u_i)\}_{i=1}^k$ weakly converge to centered Gaussian random variables $\{\Psi_R(u_i)\}_{i=1}^k$ with covariance

$$E[\Psi_R(u_i)\Psi_R(u_j)] = \frac{1}{1 + e^{u_i \vee u_j}} - \frac{6}{\pi^2} h(u_i)h(u_j), \qquad (2.6)$$

where

$$h(u) = \frac{u}{1 + e^u} + \log(1 + e^{-u}).$$

In fact, the proof of the CLT under the grand canonical ensembles is simpler than that for the canonical ensembles. The CLT under the canonical ensembles was shown by Pittel [194] Theorem 5 in the U-case and Yakubovich in the RU-case as we have cited above. Freiman, Vershik, and Yakubovich [85] proved the local CLT in the RU-case. Vershik and Yakubovich [213] studied the U-case with constraint on heights. Beltoft, Boutillier, and Enriquez [20] studied the U-case in a rectangular box. We point out that $u_i \vee u_j$ in the formula (2.6) coincides with that in $C_R(u, v)$ given in Theorem 2.2 and Theorem 2 of [85] , though it is written as $u_i \wedge u_j$ in [221]. See also Vershik and Yakubovich [214] for the fluctuation of p_1 (the maximal summand). The limit is the Gumbel distribution.

2.1.4 Large Deviation Principle

The large deviation principle (LDP) corresponding to the LLN established in Theorem 2.1 is not difficult to show under the grand canonical ensembles. Indeed, this is a simple consequence of the LDP for i.i.d. random sequences:

Theorem 2.4. *The LDP holds for $\tilde{\psi}^N$ under $\mu_U^{\varepsilon(N)}$ and $\mu_R^{\varepsilon(N)}$ with speed N and rate functions*

$$\mathbb{F}_U(\varphi) = -\int_{\mathbb{R}_+^\circ} h_U(-\varphi'(u))du + \alpha \int_{\mathbb{R}_+^\circ} \varphi(u)du + \alpha, \quad \varphi = \{\varphi(u); u \in \mathbb{R}_+^\circ\},$$

$$\mathbb{F}_R(\varphi) = -\int_{\mathbb{R}_+} h_R(-\varphi'(u))du + \beta \int_{\mathbb{R}_+} \varphi(u)du + \beta, \quad \varphi = \{\varphi(u); u \in \mathbb{R}_+\},$$

where

$$h_U(x) = -x \log x + (1+x) \log(1+x), \quad x \geq 0,$$

$$h_R(x) = -x \log x - (1-x) \log(1-x), \quad x \in [0, 1],$$

respectively. Here, the LDP for $\tilde{\psi}^N$ under $\mu_U^{\varepsilon(N)}$ with speed N and the rate function $\mathbb{F}_U$ roughly says that

$$\mu_U^{\varepsilon(N)}(\tilde{\psi}^N \sim \varphi) \sim e^{-N\mathbb{F}_U(\varphi)},$$

as $N \to \infty$. See, for example, Dembo, Vershik, and Zeitouni [61] for the precise setting and statement of the LDP.

Outline of the proof. We first consider the U-case. If we introduce the height differences $\xi_k := \psi(k-1) - \psi(k)$, $k \in \mathbb{N}$, then we see that, under μ_U^ε, $\{\xi_k\}_{k \in \mathbb{N}}$ is an independent sequence and each ξ_k has a geometric distribution with mean $\rho^\varepsilon(k)$. The mean $\rho^{\varepsilon(N)}([\varepsilon u])$ at the macroscopic point $u \in \mathbb{R}_+^\circ$ converges to the slope $\rho_U(u) := -\psi'_U(u)$ of the Vershik curve. Thus, by a similar argument to that used to show the sample path LDP (cf. Mogul'skii [182] and Dembo and Zeitouni [62]), we see that the LDP holds for $\tilde{\psi}^N$ under $\mu_U^{\varepsilon(N)}$ with the rate function

$$\mathbb{F}_U(\varphi) = \int_{\mathbb{R}_+^\circ} H(\nu_{\rho(u)}|\nu_{\rho_U(u)})du, \tag{2.7}$$

where $\rho(u) := -\varphi'(u)$,

$$H(\nu_1|\nu_2) = \sum_{\ell \in \mathbb{Z}_+} \nu_1(\ell) \log \frac{\nu_1(\ell)}{\nu_2(\ell)},$$

is the relative entropy, and ν_ρ stands for the geometric distribution with mean ρ. One can expect to have $\nu_{\rho_U(u)}$ in (2.7) because the geometric distribution with mean $\rho^{\varepsilon(N)}([\varepsilon u])$ converges to that with mean $\rho_U(u)$. Then, it is easy to see that this $\mathbb{F}_U$ coincides with that in the statement of Theorem 2.4 by noting that $\rho_U(u) = 1/(e^{\alpha u} - 1)$ and

$$\int_{\mathbb{R}_+^\circ} \left(\log(e^{\alpha u} - 1) - \alpha u \right) du = \frac{1}{\alpha} \int_0^1 \frac{1}{x} \log(1-x) dx = -\alpha.$$

The RU-case is similar. Indeed, if we consider the height differences $\eta_k := \psi(k-1) - \psi(k), k \in \mathbb{N}$, then we see that, under μ_R^ε, $\{\eta_k\}_{k \in \mathbb{N}}$ is an independent sequence and each $\eta_k \in \{0, 1\}$ has a distribution with mean $\rho^\varepsilon(k)$ determined by the relations

$$\rho^\varepsilon(k+1) = \frac{\varepsilon \rho^\varepsilon(k)}{1 - (1-\varepsilon)\rho^\varepsilon(k)}, \quad \rho^\varepsilon(0) = \frac{\varepsilon}{1+\varepsilon}.$$

The mean $\rho^{\varepsilon(N)}([\varepsilon u])$ at the macroscopic point $u \in \mathbb{R}_+^\circ$ converges to the slope $\rho_R(u) := -\psi_R'(u)$ of the Vershik curve. Thus, similarly to the U-case, we see that the LDP holds for $\tilde\psi^N$ under $\mu_R^{\varepsilon(N)}$ with the rate function

$$\mathbb{F}_R(\varphi) = \int_{\mathbb{R}_+} H(\nu_{\rho(u)} | \nu_{\rho_R(u)}) du, \tag{2.8}$$

where $\rho(u) := -\varphi'(u)$ and ν_ρ stands for the distribution on $\{0, 1\}$ with mean ρ. One can easily see that this $\mathbb{F}_R$ coincides with that in the statement of Theorem 2.4 by noting that $\rho_R(u) = 1/(e^{\beta u} + 1)$ and

$$\int_{\mathbb{R}_+} \left(\log(e^{\beta u} + 1) - \beta u \right) du = \frac{1}{\beta} \int_0^1 \frac{1}{x} \log(1+x) dx = \beta. \qquad \square$$

Note that $\mathbb{F}_U(\varphi) = 0$ (the minimizer of $\mathbb{F}_U$) is equivalent to $\varphi = \psi_U$ (the limit of the LLN) and $\mathbb{F}_R(\varphi) = 0$ is equivalent to $\varphi = \psi_R$. In fact, (2.7) and (2.8) imply that $\mathbb{F}_U(\varphi) = 0$ iff $\varphi' = \psi_U'$ and $\mathbb{F}_R(\varphi) = 0$ iff $\varphi' = \psi_R'$, respectively. Since $\varphi(+\infty) = \psi_U(+\infty) = \psi_R(+\infty) = 0$, we see that these imply $\varphi = \psi_U$ and $\varphi = \psi_R$, respectively.

Dembo, Vershik, and Zeitouni [61] studied the LDP under the canonical ensembles $\mu_U^{N^2}$ and $\mu_R^{N^2}$ and showed that the rate functions are given by the above $\mathbb{F}_U(\varphi)$ with the term $\int_{\mathbb{R}_+^\circ} \varphi(u) du$ replaced by 1 in the U-case and by $\mathbb{F}_R(\varphi)$ with the term $\int_{\mathbb{R}_+} \varphi(u) du$ replaced by 1 in the RU-case, since the macroscopic area of the scaled Young diagrams is always 1 (though this does not provide proof for the LDP under the canonical ensembles).

2.1.5 Equivalence of Ensembles under Inhomogeneous Conditioning

We now consider the canonical ensemble μ_R^M on $\mathcal{Q}_M$ in the RU-case with further conditioning on its size, i.e., on its maximal summand q_1 such that $q_1 \leq L$ and its support K determined by $q_K = 1$ and $q_i = 0$ for $i \geq K+1$ (the latter automatically holds from $q_K = 1$ and $q \in \mathcal{D}_M$). More precisely, we consider the uniform probability measure $\mu_{K,M,L}$ on the set $\mathcal{D}_{K,M,L} = \{q = (q_i)_{i \in \mathbb{N}} \in \mathcal{D}_M; L \geq q_1 > q_2 > \cdots > q_K = 1\}$ for $K, M, L \in \mathbb{N}$. The height function $\psi_q = \{\psi_q(u); u \in \mathbb{R}_+\}$ of the Young diagram corresponding to $q \in \mathcal{D}_{K,M,L}$ is defined by (2.1). Then, from $q \in \mathcal{D}_{K,M,L}$, ψ_q satisfies

$$\psi_q(0) = K, \quad \int_0^\infty \psi_q(u)du = M, \quad \psi_q(u) = 0 \ \text{ for } u \geq L.$$

The goal is to study the asymptotic behavior (called the thermodynamic limit) of $\mu_{K,M,L}$ as the sizes $K, M, L \to \infty$ in such a way that $K \sim L, M \sim L^2$. This is motivated by the study of the corresponding dynamics which we will discuss later in Sect. 2.2.5.

To this end, it turns out more convenient to work with the corresponding height differences. As we already mentioned, one can identify ψ_q with its *height difference* $\eta : \mathbb{N} \to \{0, 1\}$, defined by

$$\eta_k = \psi(k-1) - \psi(k), \quad k \in \mathbb{N}. \tag{2.9}$$

In terms of $\eta = (\eta_k)_{k \in \mathbb{N}}$, $q \in \mathcal{D}_{K,M,L}$ implies that

$$\psi_q(0) = \sum_{k \in \mathbb{N}} \eta_k = K,$$

$$\int_0^\infty \psi_q(u)du = \sum_{k \in \mathbb{N}} k\eta_k = M, \tag{2.10}$$

and

$$\eta_k = 0 \ \text{ for } k \geq L+1.$$

The last condition allows us to consider η defined only on the finite interval $[1, L] \cap \mathbb{Z}$. Assuming that L is odd, $L = 2\ell + 1$ with $\ell \in \mathbb{N}$, and shifting the system by $-\ell$, one can reformulate the problem as follows: Let $\Lambda_\ell = \{-\ell, \ldots, \ell\}$ and consider a particle configuration $\eta = (\eta_k)_{k \in \Lambda_\ell} \in \{0, 1\}^{\Lambda_\ell}$ on Λ_ℓ. For $\eta \in \{0, 1\}^{\Lambda_\ell}$, we define $K_{\Lambda_\ell}(\eta)$ and $M_{\Lambda_\ell}(\eta)$ by

$$K_{\Lambda_\ell}(\eta) := \sum_{k \in \Lambda_\ell} \eta_k, \quad M_{\Lambda_\ell}(\eta) := \sum_{k \in \Lambda_\ell} k\eta_k.$$

Then, the configuration space of particles on Λ_ℓ is defined by $\Sigma_{\Lambda_\ell,K,M} = \{\eta \in \{0, 1\}^{\Lambda_\ell}; K_{\Lambda_\ell}(\eta) = K, M_{\Lambda_\ell}(\eta) = M\}$ and consider the canonical ensembles, that is, the uniform probability measures $\nu_{\Lambda_\ell,K,M}$ on $\Sigma_{\Lambda_\ell,K,M}$, see Fig. 2.6.

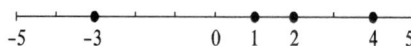

Fig. 2.6 $\ell = 5, K = 4, M = 4$

Our goal is to study the asymptotic behavior of $\nu_{\Lambda_\ell,K,M}$ as $\ell \to \infty$ with $K \sim \ell, M \sim \ell^2$. This is actually equivalent to proving the so-called *equivalence of ensembles* (i.e., the equivalence of canonical and grand canonical ensembles in the limit), or the local equilibrium for the system having two conserved quantities K_{Λ_ℓ} and M_{Λ_ℓ}. Since M_{Λ_ℓ} is spatially inhomogeneous, the limit profile is also inhomogeneous. This profile transformed back to the level of height functions is related to the Vershik curve, see Sect. 2.1.6.

To formulate the result, we introduce the grand canonical ensembles, that is, the Bernoulli measures ν_α on $\Sigma := \{0, 1\}^{\mathbb{Z}}$ with mean $\alpha, \alpha \in (0, 1)$.

The following theorem is shown by [106]. For the proof, we extend the classical local limit theorem for a sum of Bernoulli independent sequences to those multiplied by linearly growing weights.

Theorem 2.5. *Let $K = K_\ell, M = M_\ell, k_j = k_{\ell,j}$ be given for $\ell \in \mathbb{N}, 1 \le j \le p$ and satisfy*

$$\lim_{\ell \to \infty} \frac{K}{2\ell + 1} = \rho \in (0, 1),$$

$$\lim_{\ell \to \infty} \frac{M}{(2\ell + 1)^2} = m \in \left(-\frac{1}{2}\rho(1 - \rho), \frac{1}{2}\rho(1 - \rho) \right),$$

$$\lim_{\ell \to \infty} \frac{k_j}{\ell} = x_j \in (-1, 1).$$

Then, if $\{x_j\}_{j=1}^p$ are distinct, for all local functions $f_j = f_j(\eta)$ on $\Sigma, 1 \le j \le p$, we have that

$$\lim_{\ell \to \infty} E_{\nu_{\Lambda_\ell,K,M}} \left[\prod_{j=1}^{p} \tau_{k_j} f_j \right] = \prod_{j=1}^{p} E_{\nu_{\beta(x_j)}}[f_j], \tag{2.11}$$

where τ_k are shifts on Σ defined by $\tau_k \eta = (\eta_{i+k})_{i \in \mathbb{Z}} \in \Sigma$ for $\eta \in \Sigma$ and $\tau_k f(\eta) = f(\tau_k \eta)$, and

$$\beta(x) \equiv \beta(x; a, b) = \frac{e^{bx} a}{e^{bx} a + (1 - a)}, \quad x \in [-1, 1],$$

with two parameters $a \in (0, 1)$ and $b \in \mathbb{R}$ determined from ρ and m by the relations

$$\frac{1}{2} \int_{-1}^{1} \beta(x; a, b)dx = \rho, \quad \frac{1}{4} \int_{-1}^{1} x\beta(x; a, b)dx = m. \tag{2.12}$$

Fig. 2.7 Scaling from height
difference η to macroscopic
profile β

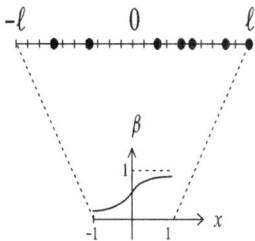

The convergence is uniform in (K, M) in the region specified by $\varepsilon \le K/(2\ell + 1) \le 1 - \varepsilon$ and $M/(2\ell + 1)^2 \in (-v/2 + \varepsilon, v/2 - \varepsilon)$ for every $\varepsilon > 0$, see Fig. 2.7.

Nagahata [189] extended Theorem 2.5 by showing an error estimate on the difference of two terms in (2.11), which is uniform in ℓ, K, M.

For every ρ and m, one can find a and b uniquely. In fact, the relation (2.12) defines a diffeomorphism

$$(a, b) \in (0, 1) \times \mathbb{R} \longmapsto (\rho, m) \in D = \left\{ 0 < \rho < 1, |m| < \frac{1}{2}\rho(1 - \rho) \right\}.$$

This theorem asserts that, as $\ell \to \infty$ under the canonical ensemble $\nu_{\Lambda_\ell, K, M}$, the limit distributions are asymptotically independent for microscopic regions which are macroscopically separated, and the microscopic limit distribution around the macroscopic point $x \in (-1, 1)$ is the grand canonical ensemble $\nu_{\beta(x)}$ with macroscopically dependent profile $\beta(x)$:

$$\lim_{\ell \to \infty, \frac{k}{\ell} \to x} \nu_{\Lambda_\ell, K, M} \circ \tau_k^{-1} = \nu_{\beta(x)}.$$

Such a situation is called *local equilibrium* in statistical physics.

Outline of the proof of Theorem 2.5: The proof is divided into four steps.

Step 1: The first observation is the following, see Lemma 4.1 of [106] for details. If $\beta(\cdot) = \beta(\cdot; a, b)$ for some a, b, then

$$\nu_{\beta(\cdot)}^{\Lambda_\ell}(\cdot | \Sigma_{\Lambda_\ell, K, M}) = \nu_{\Lambda_\ell, K, M}(\cdot),$$

where $\nu_{\beta(\cdot)}^{\Lambda_\ell}$ denotes the distribution of independent random variables $\{\eta_k\}_{k \in \Lambda_\ell}$ such that $E[\eta_k] = \beta(k/\ell)$. This explains why the functions β of the above form appear in the limit.

Step 2: To show (2.11), we consider only the case $p = 1$ for simplicity, so that k satisfies $\lim_{\ell \to \infty} \frac{k}{\ell} = x \in (-1, 1)$. From Step 1, we have that

$$E_{\nu_{\Lambda_\ell,K,M}}[\tau_k f] - E_{\nu_{\beta(x)}}[f]$$

$$= \sum_{\xi \in \{0,1\}^{\Gamma+k}} \{f(\xi) - E_{\nu_{\beta(x)}}[f]\} \frac{v_{\beta(\cdot)}^{\Lambda_\ell}(\eta|_{\Gamma+k} = \xi, K_{\Lambda_\ell}(\eta) = K, M_{\Lambda_\ell}(\eta) = M)}{v_{\beta(\cdot)}^{\Lambda_\ell}(K_{\Lambda_\ell}(\eta) = K, M_{\Lambda_\ell}(\eta) = M)},$$

for every local function f with support $\Gamma \Subset \mathbb{Z}$.

Step 3: To compute the ratio of two probabilities in the formula in Step 2, we apply
the following *local central limit theorem* for $(K_{\Lambda_\ell}(\eta), M_{\Lambda_\ell}(\eta))$ distributed under
$v_{\beta(\cdot)}^{\Lambda_\ell}$.

Proposition 2.1. *As* $\ell \to \infty$,

$$\sqrt{U_\ell V_\ell}\, v_{\beta(\cdot)}^{\Lambda_\ell}(K_{\Lambda_\ell}(\eta) = K, M_{\Lambda_\ell}(\eta) = M) = q_0(y_1, y_2) + o(1)$$

holds uniformly in K, M, where

$$y_1 = \frac{1}{\sqrt{U_\ell}}(K - E_\ell), \quad y_2 = \frac{1}{\sqrt{V_\ell}}(M - F_\ell),$$

$$E_\ell \equiv E[K_{\Lambda_\ell}] = \sum_{k \in \Lambda_\ell} \beta(k/\ell), \quad F_\ell \equiv E[M_{\Lambda_\ell}] = \sum_{k \in \Lambda_\ell} k\beta(k/\ell),$$

$$U_\ell \equiv V[K_{\Lambda_\ell}] = \sum_{k \in \Lambda_\ell} \beta(k/\ell)(1 - \beta(k/\ell)),$$

$$V_\ell \equiv V[M_{\Lambda_\ell}] = \sum_{k \in \Lambda_\ell} k^2 \beta(k/\ell)(1 - \beta(k/\ell)),$$

$$q_0(y_1, y_2) = \frac{1}{2\pi\sqrt{1 - \lambda^2}} \exp\left\{-\frac{y_1^2 - 2\lambda y_1 y_2 + y_2^2}{2(1 - \lambda^2)}\right\},$$

with

$$\lambda = \frac{\int_{-1}^{1} x\beta(x)(1 - \beta(x))dx}{\sqrt{\int_{-1}^{1} \beta(x)(1 - \beta(x))dx \int_{-1}^{1} x^2\beta(x)(1 - \beta(x))dx}}, \quad |\lambda| < 1.$$

Proof of Proposition 2.1. We need to study the sums of independent random variables

$$K_{\Lambda_\ell} = \sum_{k \in \Lambda_\ell} \eta_k \quad \text{and} \quad M_{\Lambda_\ell} = \sum_{k \in \Lambda_\ell} k\eta_k,$$

with growing weights k. In particular, $\{k\eta_k\}_k$ doesn't satisfy "good" moment conditions required for the proof of the classical local limit theorem, cf. Petrov [192], so that we need to extend and modify the argument in [192] to fit our setting. We rewrite as follows:

Fig. 2.8 Estimate for $f_\ell(s, t)$

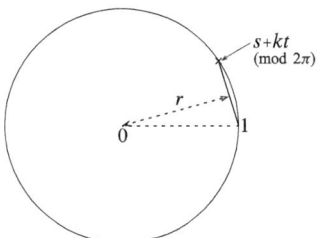

$$(2\pi)^2 v_{\beta(\cdot)}^{\Lambda_\ell}\left(K_{\Lambda_\ell} = K, M_{\Lambda_\ell} = M\right) = \int_{-\pi}^{\pi}\int_{-\pi}^{\pi} e^{-i(sK + tM)} f_\ell(s, t)\, ds dt,$$

where

$$f_\ell(s, t) := E\left[e^{i(sK_{\Lambda_\ell} + tM_{\Lambda_\ell})}\right].$$

Then, we divide the region $[-\pi, \pi]^2$ of the integration in the right-hand side into three parts. One of them is

$$I = \int_{c \leq |s| \leq \pi, \frac{c}{\ell} \leq |t| \leq \pi} e^{-i(sK + tM)} f_\ell(s, t)\, ds dt,$$

with some small $c > 0$. The exponential decay of I as $\ell \to \infty$ follows by showing

$$|f_\ell(s, t)| = \prod_{k \in \Lambda_\ell}\left|\beta(k/\ell)e^{i(s+kt)} + (1 - \beta(k/\ell))\right| \leq Cr^{p\ell},$$

with some $0 < r, p < 1$ and $C > 0$, for $c \leq |s| \leq \pi, \frac{c}{\ell} \leq |t| \leq \pi$, see Fig. 2.8. We omit estimates of the other two parts, see [106] Proposition 3.1 for details. $\square$

Step 4: We continue the proof of Theorem 2.5. If we choose $\beta(\cdot)$ in such a way that y_1 and y_2 as appeared in Proposition 2.1 are both close to 0, then, from Proposition 2.1 with $q_0(y_1, y_2)$ close to 1, we see that the ratio of the two probabilities appearing in Step 2 converges to $v_{\beta(x)}(\xi)$ and this leads to (2.11) when $p = 1$. The first relation in (2.12) follows from $K \sim E_\ell$, $K \sim 2\ell\rho$, $E_\ell \sim \ell\int_{-1}^{1}\beta(x)dx$ as $\ell \to \infty$, while the second one follows from $M \sim F_\ell$, $M \sim 4\ell^2 m$, $F_\ell \sim \ell^2\int_{-1}^{1} x\beta(x)dx$ as $\ell \to \infty$. $\square$

2.1.6 Related Young Diagrams

The macroscopic profile $\beta(x)$ provided by Theorem 2.5 has a connection to the Vershik curve $y = \psi_R(u)$ appearing in Theorem 2.1 in a scaling limit under the

Fig. 2.9 Height function ψ^ℓ

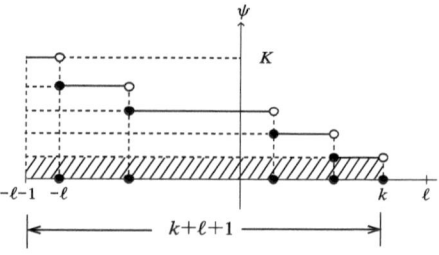

RU-statistics. In fact, one can associate the height function $\psi^\ell(u), u \in [-\ell - 1, \ell]$ of the Young diagram with the particle configuration $\eta \in \Sigma_\ell = \{0, 1\}^{\Lambda_\ell}$ by the rule

$$\psi^\ell(u) = \sum_{k \in \Lambda_\ell : k > u} \eta_k, \quad u \in [-\ell - 1, \ell], \tag{2.13}$$

see Fig. 2.9. Note that ψ^ℓ is a right-continuous non-increasing step function and satisfies

$$\psi^\ell(-\ell - 1) = K_{\Lambda_\ell}(\eta), \quad \psi^\ell(\ell) = 0, \tag{2.14}$$

i.e., the height and the side length of the associated Young diagram are K_{Λ_ℓ} and $2\ell + 1$, respectively, with the area

$$\int_{-\ell-1}^{\ell} \psi^\ell(u) \, du = (\ell + 1)K_{\Lambda_\ell}(\eta) + M_{\Lambda_\ell}(\eta). \tag{2.15}$$

Under the distribution $\nu_{\Lambda_\ell, K, M}$, we consider the macroscopically scaled height function defined by

$$\tilde{\psi}^\ell(x) := \frac{1}{\ell} \psi^\ell(\ell x), \quad x \in [-1, 1]. \tag{2.16}$$

Corollary 2.1. *Under the assumptions of Theorem 2.5, $\tilde{\psi}^\ell$ converges as $\ell \to \infty$ to ψ in probability in the following sense:*

$$\lim_{\ell \to \infty} \nu_{\Lambda_\ell, K, M} \left(\sup_{x \in [-1,1]} |\tilde{\psi}^\ell(x) - \psi(x)| > \delta \right) = 0,$$

for every $\delta > 0$. The limit ψ is defined by $\psi(x) = \int_x^1 \beta(y)dy$, $x \in [-1, 1]$ with $\beta(x)$ determined in Theorem 2.5, see Fig. 2.10.

Fig. 2.10 The limit ψ

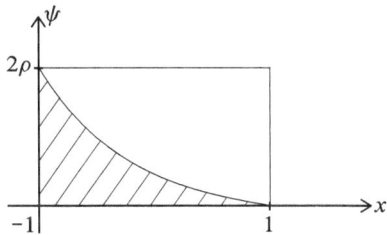

The limit ψ has the slope $\psi'(x) = -\beta(x)$ and satisfies

$$\psi(-1) = 2\rho, \quad \psi(1) = 0, \quad \int_{-1}^{1} \psi(x)dx = 2\rho + 4m,$$

and the ordinary differential equation (ODE)

$$\psi'' + c\psi'(1 + \psi') = 0,$$

with $c = -b$. Note that the Vershik curve $y = \psi_R(u)$ in the RU-case satisfies the same ODE with $c = \pi/\sqrt{12}$. This is also seen from the fact that it is a stationary solution of the hydrodynamic equation (2.26) obtained in Sect. 2.2.2 below. Note that this implies that $\beta = -\psi'$ is a stationary solution of the viscous Burgers equation $\partial_t\beta = \beta'' - b(\beta(1-\beta))'$.

In Beltoft, Boutillier, and Enriquez [20] (and Pittel [194]), the grand canonical ensembles (and canonical ensembles, respectively) in a rectangular box for the U-statistics are treated using combinatorial methods, while here we have discussed the canonical ensembles in a rectangular box for the RU-statistics via a probabilistic approach. In Vershik and Yakubovich [213], the limit curves together with fluctuations are studied under the grand canonical and canonical ensembles for the U-statistics.

2.2   Dynamic Theory of 2D Young Diagrams

So far, we have discussed the static theory for random 2D Young diagrams and studied their scaling limits under U/RU and canonical/grand canonical ensembles. Now we turn to the corresponding dynamic theory. We will mostly discuss non-conservative systems associated with the grand canonical ensembles except for Sect. 2.2.5, where we discuss conservative systems associated with the canonical ensembles. The purpose is to study and extend the results of the static theory from a dynamical point of view.

In Sect. 2.2.1, we construct dynamics of 2D Young diagrams associated with the U- and RU-grand canonical ensembles, by allowing the creation and annihilation of unit squares located at the boundary (interface) of the diagrams. We then show in Sect. 2.2.2 that, as the averaged size of the diagrams diverges, the corresponding macroscopic height variable converges to a solution of a certain non-linear partial differential equation under a proper hydrodynamic scaling. Furthermore, the stationary solution of the limit equation is identified with the Vershik curve. This gives a dynamical proof for the derivation of the Vershik curves. Section 2.2.3 discusses the non-equilibrium fluctuation problem, which corresponds to the hydrodynamic limit, and derives linear stochastic partial differential equations in the limit. We show that their invariant measures are identical to the Gaussian measures appeared in Theorem 2.2. Section 2.2.4 discusses the dynamic LDP.

2.2.1 Dynamics of 2D Young Diagrams and Their Gradient Fields

To the U- and RU- grand canonical ensembles, one can associate a weakly asymmetric zero-range process (WAZRP) p_t on $\mathscr{P}$, respectively a weakly asymmetric simple exclusion process (WASEP) q_t on $\mathscr{Q}$, with a stochastic reservoir at the boundary site $\{0\}$ in both processes as natural time evolutions of the Young diagrams, or more precisely, those of the gradients of their height functions.

We first consider the U-case and construct the dynamics of 2D Young diagrams, which have the U-grand canonical ensemble μ_U^ε defined in (2.2) as their invariant measures. Let $p_t \equiv p_t^\varepsilon = (p_i(t))_{i \in \mathbb{N}}$ be the Markov process on $\mathscr{P}$ defined by means of the infinitesimal generator $L_{\varepsilon,U}$ acting on functions $f : \mathscr{P} \to \mathbb{R}$ as

$$L_{\varepsilon,U}f(p) = \sum_{i \in \mathbb{N}} \left[\varepsilon 1_{\{p_{i-1} > p_i\}} \{f(p^{i,+}) - f(p)\} + 1_{\{p_i > p_{i+1}\}} \{f(p^{i,-}) - f(p)\} \right], \quad (2.17)$$

where $p^{i,\pm} = (p_j^{i,\pm})_{j \in \mathbb{N}} \in \mathscr{P}$ are defined by

$$p_j^{i,\pm} = \begin{cases} p_j, & \text{if } j \neq i, \\ p_i \pm 1, & \text{if } j = i. \end{cases} \quad (2.18)$$

In (2.17), we put $p_0 = \infty$. Note that $n(p_t)\left(= \sum_{i \in \mathbb{N}} p_i(t)\right)$ and $\mathfrak{n}(p_t) := \sharp\{i \in \mathbb{N}; p_i(t) \geq 1\}$ change in time, but always stay finite. We set $\mathfrak{n}(p_t) = 0$ if $p_t \in \mathscr{P}_0$. It is easy to see that μ_U^ε is invariant under such dynamics for every $0 < \varepsilon < 1$ by showing that $\sum_{p \in \mathscr{P}} L_{\varepsilon,U}f(p)\mu_U^\varepsilon(p) = 0$ for a sufficiently wide class of functions f.

The evolution p_t on $\mathscr{P}$ determines an evolution of the height functions ψ_{p_t} of 2D Young diagrams through the map $p \mapsto \psi_p$ defined by (2.1), see Fig. 2.11.

We can introduce an equivalent system of particles on $\mathbb{N}$, namely we think of $p_i(t)$ as the position of the ith particle. The total number of particles $\mathfrak{n}(p_t)$ on the region $\mathbb{N}$ changes only through the creation and annihilation of particles at the boundary site $\{0\}$. In fact, the first part in the sum (2.17) with $i = \mathfrak{n}(p) + 1$ encodes the fact

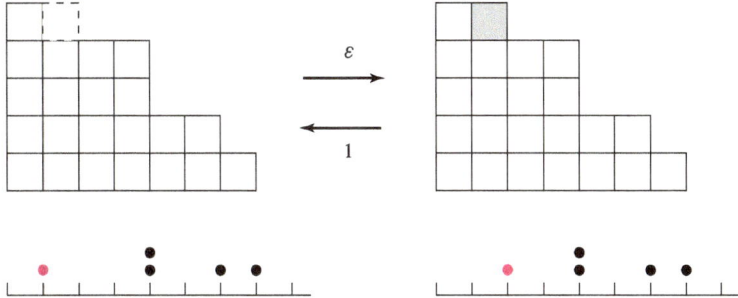

Fig. 2.11 Dynamics associated with grand canonical ensembles in the U-case

that a new particle emerges from the boundary site $\{0\}$ to the site $\{1\}$ (in $\mathbb{N}$) with rate ε, while the second part with $i = \mathfrak{n}(p)$ indicates that a particle at $\{1\}$ jumps to $\{0\}$ and disappears with rate 1. In other words, a stochastic reservoir is located at the boundary site $\{0\}$ of $\mathbb{N}$.

Next, we consider the RU-case and construct the dynamics associated with the RU-grand canonical ensemble $\mu_{\mathrm{R}}^{\varepsilon}$ defined in (2.3). Let $q_t \equiv q_t^{\varepsilon} = (q_i(t))_{i \in \mathbb{N}}$ be the Markov process on $\mathcal{Q}$ with the infinitesimal generator $L_{\varepsilon,\mathrm{R}}$ acting on functions $f : \mathcal{Q} \to \mathbb{R}$ as

$$L_{\varepsilon,\mathrm{R}}f(q) = \sum_{i \in \mathbb{N}} \left[\varepsilon 1_{\{q_{i-1}>q_i+1\}} \{f(q^{i,+}) - f(q)\} + 1_{\{q_i>q_{i+1}+1 \text{ or } q_i=1\}} \{f(q^{i,-}) - f(q)\} \right],$$

$$(2.19)$$

where $q^{i,\pm} \in \mathcal{Q}$ are defined by formula (2.18) and we put $q_0 = \infty$. It is easy to see that $\mu_{\mathrm{R}}^{\varepsilon}$ is invariant under this dynamics. Similarly to the U-case, we think of $q_i(t)$ as the position of the ith particle. The model defined by the generator (2.19) involves a stochastic reservoir at $\{0\}$. The only difference is that the creation of a new particle at $\{1\}$ is allowed if this site is vacant, since, when $q_i = 0$, the transition from q to $q^{i,+} \in \mathcal{Q}$ occurs only when $q_{i-1} \geq 2$.

As we already saw, the Young diagrams can be recovered from their height differences, i.e., gradient fields defined by $\xi(k) := \psi(k-1) - \psi(k) \in \mathbb{Z}_+$, $k \in \mathbb{N}$ in the U-case and $\eta(k) := \psi(k-1) - \psi(k) \in \{0, 1\}$, $k \in \mathbb{N}$ in the RU-case, respectively. In this description,

$$\xi_t(k) = \psi_t(k-1) - \psi_t(k) \in \mathbb{Z}_+, \quad k \in \mathbb{N},$$

and

$$\eta_t(k) = \psi_t(k-1) - \psi_t(k) \in \{0, 1\}, \quad k \in \mathbb{N}, \tag{2.20}$$

represent the number of (unlabeled) particles at the site $\{k\}$ in the U- and RU-cases, respectively. For convenience, we set $\xi_t(0) = \eta_t(0) = \infty$. The dynamics of height differences $\xi_t = \{\xi_t(k)\}_{k \in \mathbb{Z}_+}$ in the U-case and $\eta_t = \{\eta_t(k)\}_{k \in \mathbb{Z}_+}$ in the RU-case is a weakly asymmetric zero-range process (WAZRP) with weakly asymmetric stochastic reservoir at the boundary $k = 0$, and a weakly asymmetric simple exclusion process (WASEP) with weakly asymmetric stochastic reservoir at the boundary $k = 0$, respectively.

In summary, we have three different, but essentially equivalent descriptions of the dynamics of 2D Young diagrams: p_t, ψ_t, ξ_t in the U-case and q_t, ψ_t, η_t in the RU-case, respectively.

2.2.2 Hydrodynamic Limits (LLN)

The results of this section are due to Funaki and Sasada [114]. Under the diffusive scaling in space and time and choosing the parameter $\varepsilon = \varepsilon(N)$ of the grand canonical ensembles such that the averaged size of the Young diagrams is N^2 as in (2.5), we will derive the hydrodynamic equations in the limit and show that the Vershik curves defined in Theorem 2.1 are actually unique stationary solutions to the limiting non-linear partial differential equations in both the U- and RU-cases.

First, let us formulate the result in the U-case. For a probability measure ν on $\mathscr{P}$ and $N \geq 1$, we denote by $\mathbb{P}_\nu^N$ the distribution on the path space $D(\mathbb{R}_+, \mathscr{P})$ of the process $p_t \equiv p_t^N$ with generator $N^2 L_{\varepsilon_U(N), U}$, which is accelerated by the factor N^2, and the initial measure ν. Here, $\varepsilon_U(N)$ is defined by (2.5). Let X_U be the function space defined by

$$X_U := \{\psi : \mathbb{R}_+^\circ \to \mathbb{R}_+^\circ ; \; \psi \in C^1, \; \psi' < 0, \lim_{u \downarrow 0} \psi(u) = \infty, \; \lim_{u \uparrow \infty} \psi(u) = 0\},$$

where $\psi' = d\psi/du$. Recall that the scaled height variable $\tilde{\psi}_p^N(u)$ is defined by (2.4) for $p \in \mathscr{P}$. In other words, we define $\tilde{\psi}_{p_t}^N(u)$ by the diffusive scaling in space and time:

$$\tilde{\psi}_{p_t}^N(u) := \frac{1}{N} \psi_{N^2 t}(Nu), \quad u > 0,$$

starting from the process $\psi_t(u)$ determined by the generator $L_{\varepsilon_U(N), U}$ (without the factor N^2). Then, we have the following result.

Theorem 2.6. *Let* $(\nu^N)_{N \geq 1}$ *be a sequence of probability measures on* $\mathscr{P}$ *such that*

$$\lim_{N \to \infty} \nu^N \left(\left| \int_0^\infty f(u) \tilde{\psi}_p^N(u) du - \int_0^\infty f(u) \psi_0(u) du \right| > \delta \right) = 0 \qquad (2.21)$$

holds for every $\delta > 0$, $f \in C_0(\mathbb{R}_+^\circ)$ *and some function* $\psi_0 \in X_U$, *where* $C_0(\mathbb{R}_+^\circ)$ *is the class of all functions* $f \in C(\mathbb{R}_+^\circ)$ *having compact supports in* $\mathbb{R}_+^\circ$. *Then, for every* $t > 0$,

$$\lim_{N\to\infty} \mathbb{P}^N_{\nu^N}\left(\sup_{u\in[u_0,u_1]} |\tilde{\psi}^N_{p_t}(u) - \psi(t,u)| > \delta\right) = 0 \tag{2.22}$$

holds for every $\delta > 0$ *and* $0 < u_0 < u_1$, *where* $\psi(t,u) = \psi_U(t,u)$ *is the unique classical solution (in the space* X_U*) of the non-linear partial differential equation (PDE)*

$$\begin{cases} \partial_t\psi = \partial_u\left(\dfrac{\partial_u\psi}{1 - \partial_u\psi}\right) + \alpha\dfrac{\partial_u\psi}{1 - \partial_u\psi}, \\[2mm] \psi(0,\cdot) = \psi_0(\cdot), \\[2mm] \psi(t,\cdot) \in X_U, \quad t \ge 0, \end{cases} \tag{2.23}$$

where $\partial_t\psi = \partial\psi/\partial t$, $\partial_u\psi = \partial\psi/\partial u$, *and* $\alpha = \pi/\sqrt{6}$.

The function ψ_U defined in Theorem 2.1 is the unique stationary solution in the class X_U of the Eq. (2.23). In this way, the Vershik curve is derived from the dynamical point of view.

Remark 2.1. Spohn discussed in [205, Appendix A] 2D interfacial dynamics, motivated by the zero-temperature Ising model, under the periodic boundary condition with symmetric jump rates and derived the non-linear PDE (2.23) with $\alpha = 0$ under the hydrodynamic scaling limit. He studied interfaces having graphical representations as in our setting, but not necessarily being monotone.

We next discuss the RU-case. For a probability measure ν on $\mathscr{Q}$ and $N \ge 1$, we denote by $\mathbb{Q}^N_\nu$ the distribution on the path space $D(\mathbb{R}_+, \mathscr{Q})$ of the process $q_t \equiv q^N_t$ with generator $N^2 L_{\varepsilon_R(N),R}$ and the initial measure ν. Here, $\varepsilon_R(N)$ is defined by (2.5). Let X_R be the function space defined by

$$X_R := \{\psi : \mathbb{R}_+ \to \mathbb{R}_+; \psi \in C^1, -1 \le \psi' \le 0, \psi'(0) = -1/2, \lim_{u\uparrow\infty}\psi(u) = 0\}.$$

Our second theorem is stated as follows. The scaled height variable $\tilde{\psi}^N_q(u)$ is defined by (2.4) for $q \in \mathscr{Q}$.

Theorem 2.7. *Let* $(\nu^N)_{N\ge1}$ *be a sequence of probability measures on* $\mathscr{Q}$ *such that*

$$\lim_{N\to\infty} \nu^N\left(\left|\int_0^\infty f(u)\tilde{\psi}^N_q(u)du - \int_0^\infty f(u)\psi_0(u)du\right| > \delta\right) = 0 \tag{2.24}$$

holds for every $\delta > 0$, $f \in C_0(\mathbb{R}^\circ_+)$ *and some function* $\psi_0 \in X_R$. *Then, for every* $t > 0$,

$$\lim_{N\to\infty} \mathbb{Q}^N_{\nu^N}\left(\sup_{u\in[u_0,u_1]} |\tilde{\psi}^N_{q_t}(u) - \psi(t,u)| > \delta\right) = 0 \tag{2.25}$$

holds for every $\delta > 0$ and $0 < u_0 < u_1$, where $\psi(t, u) = \psi_R(t, u)$ is the unique classical solution (in the space X_R) of the non-linear partial differential equation

$$\begin{cases} \partial_t \psi = \partial_u^2 \psi + \beta \, \partial_u \psi (1 + \partial_u \psi), \\ \psi(0, \cdot) = \psi_0(\cdot), \\ \psi(t, \cdot) \in X_R, \quad t \geq 0, \end{cases} \tag{2.26}$$

and $\beta = \pi/\sqrt{12}$.

The function ψ_R defined in Theorem 2.1 is the unique stationary solution of the Eq. (2.26) in the class X_R.

Heuristic derivation of PDEs (2.23) **and** (2.26): Before addressing the proofs of Theorems 2.6 and 2.7, it might be worthy giving a heuristic derivation of these two equations. We mention only about (2.23) and don't touch the derivation of the boundary condition.

The main idea lies in the averaging effect under the local ergodicity which the system has. Namely, under the time change $t \mapsto N^2 t$, the system spends a long time and can be expected to reach equilibrium states, but locally in macroscopic time and space. This is called local equilibrium.

Let us take a test function $f \in C_0^2(\mathbb{R}_+^\circ)$. To derive (2.23), we show that $\langle \tilde{\psi}_{p_t}^N, f \rangle$ converges to $\langle \psi_t, f \rangle$ as $N \to \infty$, where $\langle \psi, f \rangle = \int_0^\infty \psi(u) f(u) du$ and ψ_t is the solution of (2.23). First, by a simple computation, we can rewrite

$$\langle \tilde{\psi}_p^N, f \rangle = \int_0^\infty \frac{1}{N} \psi_p(Nu) f(u) du$$

$$= \frac{1}{N^2} \sum_{x \in \mathbb{Z}_+} \psi_p(x) \bar{f}(x/N)$$

$$= \frac{1}{N^2} \sum_{x \in \mathbb{Z}_+} \left(\sum_{y=x+1}^\infty \xi(y) \right) \bar{f}(x/N)$$

$$= \frac{1}{N^2} \sum_{y \in \mathbb{N}} \left(\sum_{x=0}^{y-1} \bar{f}(x/N) \right) \xi(y),$$

where $\bar{f}(x/N) = N \int_{x/N}^{(x+1)/N} f(u) du$. Here, we have used the fact that ψ_p is a step function, i.e., $\psi_p(Nu) = \psi_p(x)$ for $u \in [x/N, (x+1)/N)$ for the second line, $\xi(x) = \psi_p(x-1) - \psi_p(x)$ is a height difference in the third line, and we applied the summation by parts for the last line.

The stochastic time derivative is computed as follows:

$$d\langle \tilde{\psi}^N_{p_t}, f\rangle = N^2 \frac{1}{N^2} \sum_{y\in\mathbb{N}} \left(\sum_{x=0}^{y-1} \bar{f}(x/N) \right) (\tilde{L}_\varepsilon \xi(y))(\xi^N_t) dt + dM^N_t,$$

where M^N_t is a martingale, which tends to 0 as $N \to \infty$ as we will see later in a slightly different setting in Lemma 2.2 and after, and $\tilde{L}_\varepsilon$ is the generator of the zero-range process determined by $L_{\varepsilon,\mathrm{U}}$. More precisely,

$$\tilde{L}_\varepsilon F(\xi) = \sum_{x\in\mathbb{N}} \sum_z p(z) g(\xi(x)) \{F(\xi^{x,x+z}) - F(\xi)\},$$

for $F = F(\xi)$, where $\xi^{x,y}$ is obtained from ξ after one particle at x jumps to y, i.e., $(\xi^{x,y})(w) = \xi(x) - 1$ for $w = x$, $= \xi(y) + 1$ for $w = y$ and $= \xi(w)$ otherwise; $p(z) = \varepsilon$ for $z = 1$, $= 1$ for $z = -1$ and $= 0$ otherwise; and $g(\xi(x)) = 1_{\{\xi(x)\geq 1\}}$. Therefore, computing $\tilde{L}_\varepsilon \xi(y)$, we have

$$d\langle \tilde{\psi}^N_{p_t}, f\rangle = \sum_{y\in\mathbb{N}} \left(\sum_{x=0}^{y-1} \bar{f}(x/N) \right) \left[\varepsilon 1_{\{\xi(y-1)\geq 1\}} - \varepsilon 1_{\{\xi(y)\geq 1\}} \right.$$

$$\left. + 1_{\{\xi(y+1)\geq 1\}} - 1_{\{\xi(y)\geq 1\}} \right](\xi^N_t) dt + o(1)$$

$$= \sum_{y\in\mathbb{N}} \tilde{f}(y, N, \varepsilon) 1_{\{\xi^N_t(y)\geq 1\}} + o(1),$$

where

$$\tilde{f}(y, N, \varepsilon) := \varepsilon \sum_{x=0}^{y} \bar{f}(x/N) - \varepsilon \sum_{x=0}^{y-1} \bar{f}(x/N) + \sum_{x=0}^{y-2} \bar{f}(x/N) - \sum_{x=0}^{y-1} \bar{f}(x/N)$$

$$= \varepsilon \bar{f}(y/N) - \bar{f}((y-1)/N)$$

$$= \bar{f}(y/N) - \bar{f}((y-1)/N) - \frac{\alpha}{N}\bar{f}(y/N) + o(1/N)$$

$$= \tfrac{1}{N}\{f'(y/N) - \alpha f(y/N)\} + o(1/N),$$

since $\varepsilon = 1 - (\alpha/N) + o(1/N)$ for the third line and $\bar{f}(y/N) - \bar{f}((y-1)/N) = \frac{1}{N}f'(y/N) + o(1/N)$ for the last line. Thus, we have

$$d\langle \tilde{\psi}^N_{p_t}, f\rangle = \tfrac{1}{N} \sum_{y\in\mathbb{N}} \{f'(y/N) - \alpha f(y/N)\} 1_{\{\xi^N_t(y)\geq 1\}} dt + o(1).$$

Now we need to apply the idea of local equilibrium and the local ergodicity. Integrating the above formula in t, we have the term $\int_0^\infty 1_{\{\xi_s^N(y)\geq 1\}}ds$ in the right-hand side. This is actually a long-time average viewed at a microscopic level. The time s is macroscopic, but $\xi^N(\cdot)$ viewed at microscopic level oscillates rapidly. This fluctuation is eventually averaged out due to the ergodicity. The process ξ^N, viewed microscopically in a vicinity of (y, s), can be approximated by an infinite system ξ_t defined as follows. The leading term in the generator of ξ_t is given by $\varepsilon = 1$ in the limit as $N \to \infty$, so that ξ_t behaves as a symmetric zero-range process with the generator

$$\tilde{L}F(\xi) = \sum_{x\in\mathbb{Z}} 1_{\{\xi(x)\geq 1\}} \left[\{F(\xi^{x,x+1}) - F(\xi)\} + \{F(\xi^{x,x-1}) - F(\xi)\} \right],$$

for $F : \mathbb{Z}_+^{\mathbb{Z}} \to \mathbb{R}$. This is an infinite system with the state space $\mathbb{Z}_+^{\mathbb{Z}}$.

Since translation-invariant equilibrium states of ξ_t play an important role, we first need to characterize all of them. Indeed, it is known that the translation-invariant invariant measures of $\tilde{L}$-process ξ_t are superpositions of $\{v_a\}_{a\in[0,1]}$, where v_a is a product probability measure on $\mathbb{Z}_+^{\mathbb{Z}}$ whose marginal at each site x is given by the geometric distribution:

$$v_a(\xi(x) = k) = (1 - a)a^k, \quad k \in \mathbb{Z}_+, \ x \in \mathbb{Z}.$$

In fact, it is easy to check that v_a is invariant under $\tilde{L}$ by showing $\int \tilde{L}F(\xi)v_a(d\xi) = 0$ for every local function $F : \mathbb{Z}_+^{\mathbb{Z}} \to \mathbb{R}$. This is an easy exercise and left to the readers. The converse statement: all translation-invariant invariant measures must be superpositions of v_a, is also known, see [11, 168]. The average number of particles at each site under v_a is given by

$$\rho = E^{v_a}[\xi(0)] = \sum_{k=0}^{\infty} k(1 - a)a^k = \frac{a}{1 - a} \in [0, \infty).$$

The inverse function of $\rho = \rho(a)$ is

$$a = \frac{\rho}{1 + \rho}.$$

As we have mentioned, the distribution of $\xi_s^N(y)$ is expected to be close to one of the equilibrium states v_a (which is ergodic under spatial translation). The problem is how to determine a. If one can show that $\tilde{\psi}_{p_s}^N$ converges to ψ_s, since ξ is a height difference of ψ_p with negative sign, one can expect that the average of $\xi(y)$ is close to $-\psi_s'(y/N)$: $\rho \sim -\psi_s'(y/N)$. In other words, one can expect that the local average of $1_{\{\xi_s^N(y)\geq 1\}}$ in s could be replaced by the ensemble average of $1_{\{\xi(0)\geq 1\}}$ under v_a with $\rho = \rho(a) = -\psi_s'(y/N)$. This is the local ergodicity. Since

$$E^{\nu_\rho}[1_{\{\xi(0)\geq 1\}}] = \nu_\rho(\xi(0) \geq 1) = \sum_{k\geq 1}(1-a)a^k = a,$$

under the time average, one can replace $1_{\{\xi_s^N(y)\geq 1\}}$ by $a = \frac{\rho}{1+\rho}$ with $\rho = -\psi_s'(y/N)$. Therefore, the limit $\psi_t = \psi(t, \cdot)$, if exists, would satisfy

$$\frac{d}{dt}\langle\psi_t, f\rangle = \int_0^\infty \left(f'(u) - \alpha f(u)\right)\frac{-\partial_u\psi_t}{1-\partial_u\psi_t}\,du$$

$$= \int_0^\infty \left[\partial_u\left\{\frac{\partial_u\psi_t}{1-\partial_u\psi_t}\right\} + \alpha\frac{\partial_u\psi_t}{1-\partial_u\psi_t}\right]f(u)\,du.$$

This is nothing but the weak form of (2.23).

Similar arguments work for the derivation of the Eq. (2.26). Note that the negative derivative (tilt or slope) $\rho(t, u) = -\partial_u\psi(t, u)$ of the solution of (2.26) is a classical solution of the viscous Burgers equation

$$\partial_t\rho = \partial_u^2\rho + \beta\partial_u\big(\rho(1-\rho)\big). \tag{2.27}$$

It is well-known that this equation can be derived under the diffusive hydrodynamic scaling limit from the weakly asymmetric simple exclusion process on $\mathbb{Z}$ (without boundary condition), see Gärtner [121], De Masi, Presutti, and Scacciatelli [60], and others.

Outline of the proof of Theorems 2.6 and 2.7: A known general method for attacking the problem of the hydrodynamic limit (for gradient system) is the entropy method due to Guo, Papanicolaou, and Varadhan [134], see also [168]. However, some technical difficulties exist when dealing with systems on an infinite domain (since the relative entropy diverges) or with boundary conditions, see Fritz [87] or Eyink, Lebowitz, and Spohn [76] dealing with these difficulties. If we work on a (discrete and large) torus, these difficulties are avoided.

Our system is defined on an infinite domain with boundary condition (reservoir). However, as we saw above, the limit Eq. (2.27) is the viscous Burgers equation and can be recast as a simple linear heat equation using the *Cole-Hopf transformation*. This transformation is applicable even for microscopic system and, indeed, Gärtner [121] proved the hydrodynamic limit employing this special feature of the model before the entropy method was invented. We essentially adopt his method. The key roles will be played by the *Russian transformation* and the Cole-Hopf transformation, both at the microscopic level.

We first give the proof of Theorem 2.6. The Russian transformation leads from the weakly asymmetric zero-range process with stochastic reservoir at the boundary point 0 to the weakly asymmetric simple exclusion process on the whole region $\mathbb{Z}$ without boundary. Then, one can apply the result of Gärtner [121]. The idea of transforming p_t into $\bar{\eta}_t$, which is indeed known in the study of particle systems, is

Fig. 2.12 Russian
transformation

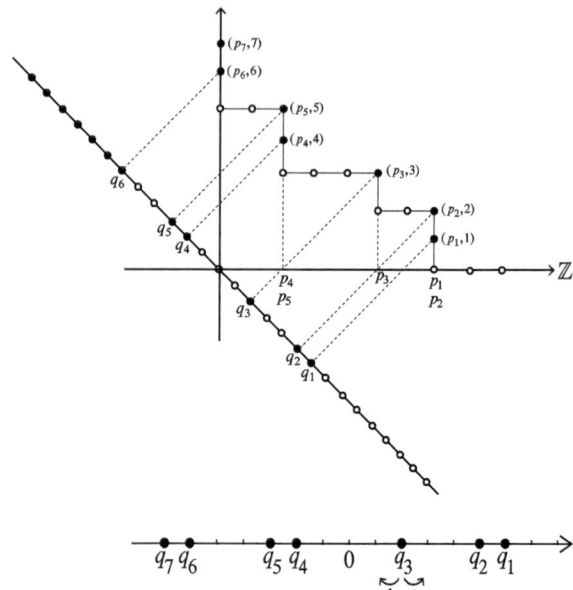

Fig. 2.13 WASEP

useful in avoiding the difficulty of treating singularities at the boundary $u = 0$,
which appear in the limit of $\tilde{\psi}^N_{p_t}(u)$.

1. *Russian transformation at microscopic level.* The name comes from the fact that
Russian researchers like Kerov [167] and Vershik frequently use this transformation.
The equivalent transformation applied for particle systems was also known and used
by Spohn and others.

Let $p = \{p_1 \geq p_2 \geq \cdots\} \in \mathscr{P}$ be given. Then, p is transformed into $q = \{q_i = p_i - i\}_{i=1}^{\infty}$. This q is obtained from p via the following transformation shown
in Fig. 2.12. We consider the two-dimensional lattice $\mathbb{Z}^2$ and place particles at (p_i, i).
Then, project these points on the line $y = -x$. Namely, map

$$(p_i, i) \longmapsto \left(\tfrac{1}{2}(p_i - i), -\tfrac{1}{2}(p_i - i)\right).$$

In fact, the condition $(a, -a) \perp (p_i, i) - (a, -a)$ implies $a = \tfrac{1}{2}(p_i - i)$. In other
words, we rotate the $\mathbb{Z}^2$-plane by 45 degree and project particles to the x-axis, and
finally scale the x-axis by $\sqrt{2}$. Then, p_i is transformed into $q_i = p_i - i$. In Fig. 2.12,
$p_1 = p_2 = 7$, $p_3 = 5$, $p_4 = p_5 = 2$, $p_6 = \cdots = 0$ are transformed to $q_1 = 6$, $q_2 = 5$, $q_3 = 2$, $q_4 = -2$, $q_5 = -3$, $q_6 = -6, \ldots$, see Fig. 2.13.

Under this map, the process p_t^ε with the generator $L_{\varepsilon, U}$ is transformed into the
weakly asymmetric simple exclusion process on $\mathbb{Z}$ with jump rates ε to the right
and 1 to the left. The configuration of such system is described by

$$\bar{\eta}(x) := \sharp\{i : q_i = x\} \in \{0, 1\}.$$

2. *Russian transformation at macroscopic level.* We consider the class of functions ρ:

$$Y_U = \left\{ \rho \in C(\mathbb{R}, (0, \infty)); \int_{-\infty}^{0} (1 - \rho(v)) dv = \int_{0}^{\infty} \rho(v) dv < \infty \right\}.$$

Recalling the class X_U of functions ψ, we define a map $\Phi_U : X_U \to Y_U$ by

$$\Phi_U(\psi)(v) = \frac{-\psi'(G_\psi^{-1}(v))}{1 - \psi'(G_\psi^{-1}(v))}, \quad v \in \mathbb{R},$$

where $G_\psi(u) = u - \psi(u)$, which is increasing in u; note that this function appears in the limit PDE (2.23). The map Φ_U defined in this way is actually the macroscopic correspondent of the Russian transformation. Namely, if a sequence $\{p \equiv p^N = (p_i)\}_{N=1}^{\infty}$ is given and has a limit

$$\psi(u) = \lim_{N \to \infty} \frac{1}{N} \sum_{i \in \mathbb{N}} 1_{\{u < \frac{p_i}{N}\}}, \tag{2.28}$$

then the limit

$$\rho(v) dv = \lim_{N \to \infty} \frac{1}{N} \sum_{i \in \mathbb{N}} \delta_{\frac{q_i}{N}}(dv),$$

with $q_i = p_i - i$, satisfies $\rho(v) = \Phi_U(\psi)(v)$. In fact, in Fig. 2.14, since $(\frac{p_i}{N}, \frac{i}{N}) \sim (u, \psi(u))$ implies $\frac{i}{N} \sim \psi(\frac{p_i}{N})$, we have for every test function $f \in C_0(\mathbb{R})$:

$$\int_{\mathbb{R}} f(v)\rho(v) dv = \lim_{N \to \infty} \frac{1}{N} \sum_{i \in \mathbb{N}} f(q_i/N)$$

$$= \lim_{N \to \infty} \frac{1}{N} \sum_{i \in \mathbb{N}} f\left(G_\psi(p_i/N)\right)$$

$$= \int_{\mathbb{R}_+} f(G_\psi(u))(-\psi'(u)) du$$

$$= \int_{\mathbb{R}} f(v)\Phi_U(\psi)(v) dv.$$

This implies $\rho(v) = \Phi_U(\psi)(v)$. In the above computation, the second equality follows from $\frac{q_i}{N} = \frac{p_i - i}{N} = \frac{p_i}{N} - \frac{i}{N} \sim \frac{p_i}{N} - \psi(\frac{p_i}{N}) = G_\psi(\frac{p_i}{N})$, the third from (2.28)

Fig. 2.14 Macroscopic
Russian transformation

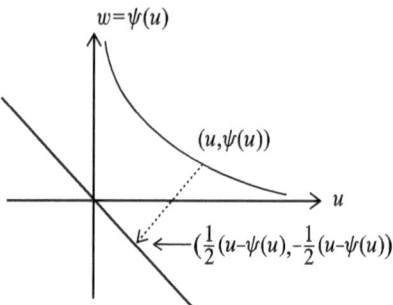

which implies $-\psi'(u)du = \lim_{N\to\infty} \frac{1}{N} \sum_{i\in\mathbb{N}} \delta_{\frac{p_i}{N}}(du)$, and the fourth by the change
of variables $v := G_\psi(u)$, which yields

$$du = \frac{du}{dv}dv = \frac{dv}{1-\psi'(u)} = \frac{dv}{1-\psi'(G_\psi^{-1}(v))}.$$

Lemma 2.1. *The map $\Phi_U : X_U \to Y_U$ is bijective. In fact, its inverse map $\Psi_U :$
$Y_U \to X_U$ is constructed as follows. For $\rho \in Y_U$, set*

$$\zeta_\rho^-(v) := \int_{-\infty}^{v} \left(1 - \rho(v')\right)dv',$$

$$\zeta_\rho^+(v) := \int_{v}^{\infty} \rho(v')dv'.$$

Then, $\zeta_\rho^- > 0$ is strictly increasing, while $\zeta_\rho^+ > 0$ is strictly decreasing, and

$$\Psi_U(\rho)(u) = \zeta_\rho^+\left((\zeta_\rho^-)^{-1}(u)\right), \quad u \in \mathbb{R}_+^\circ.$$

Proof. For a given $\rho \in Y_U$, we can construct $\psi \in X_U$ such that $\rho = \Phi_U(\psi)$. Here,
assuming that such ψ exists, we show that $\Psi_U(\rho) = \psi$, i.e., $\Psi_U \circ \Phi_U = $ id. Indeed,
since $\rho(v) = \Phi_U(\psi)(v)$ as we have seen above,

$$\zeta_\rho^-(v) = \int_{-\infty}^{v} \left(1 - \Phi_U(\psi)(v')\right)dv'$$

$$= \int_{-\infty}^{v} \left(1 + \frac{\psi'(G_\psi^{-1}(v'))}{1-\psi'(G_\psi^{-1}(v'))}\right)dv'$$

$$= \int_{0}^{G_\psi^{-1}(v)} \left(1 + \frac{\psi'(u)}{1-\psi'(u)}\right)\left(1-\psi'(u)\right)du = G_\psi^{-1}(v),$$

where the third line is obtained by the change of variables: $v' = G_\psi(u)$. This implies that $(\zeta_\rho^-)^{-1}(u) = G_\psi(u)$, and therefore

$$\Psi_U(\rho)(u) = \zeta_\rho^+\left((\zeta_\rho^-)^{-1}(u)\right) = \zeta_\rho^+\left(G_\psi(u)\right)$$

$$= \int_{G_\psi(u)}^\infty \frac{-\psi'(G_\psi^{-1}(v'))}{1 - \psi'(G_\psi^{-1}(v'))} dv'$$

$$= \int_u^\infty \frac{-\psi'(u')}{1 - \psi'(u')}\left(1 - \psi'(u')\right) du' = \psi(u).$$

This shows $\Psi_U \circ \Phi_U = \mathrm{id}$.   □

3. *Proof of Theorem 2.6.*   As we have pointed out, the process q_t deduced from p_t via the Russian transformation is the asymmetric simple exclusion process on $\mathbb{Z}$ without boundary condition and it is known that the macroscopic equation obtained under the hydrodynamic limit is the viscous Burgers equation

$$\partial_t \rho = \partial_v^2 \rho + \alpha \partial_v\left(\rho(1 - \rho)\right), \quad v \in \mathbb{R}. \tag{2.29}$$

Noting that (2.29) and (2.23) are equivalent under the map $\psi(t, u) = \Psi_U\left(\rho(t, \cdot)\right)(u)$ or $\rho(t, v) = \Phi_U\left(\psi(t, \cdot)\right)(v)$, Theorem 2.6 is shown.

We are now in the position to give the proof of Theorem 2.7. Our method is to apply the Cole-Hopf transformation for the microscopic process q_t, which was originally introduced by Gärtner [121], see also Bertini and Giacomin [27] (applied for the KPZ equation in Chap. 5). This transformation linearizes the leading term in the time evolution q_t even at the microscopic level, so that one can avoid to show the one-block and two-block estimates (due to Guo, Papanicolaou, and Varadhan [134]), which are usually required in the procedure establishing the hydrodynamic limit. The only task left is to study the boundary behavior of the transformed process, but a rather simple argument leads to the desired ergodic property of our process at the boundary, see Lemma 2.3 below.

The process of the height difference $\eta_t(x) = \psi_{q_t}(x - 1) - \psi_{q_t}(x) \in \{0, 1\}, x \in \mathbb{N}$ is a weakly asymmetric simple exclusion process with a stochastic reservoir at 0, where q_t is the process with the generator $N^2 L_{\varepsilon_R(N), R}$. To complete the proof of Theorem 2.7, we may prove the hydrodynamic limit for η_t scaled in space instead of $\tilde{\psi}_{q_t}^N$ and show that the limit $\rho(t, u)$ satisfies the viscous Burgers equation with Dirichlet boundary condition at $u = 0$

$$\begin{cases} \partial_t \rho = \partial_u^2 \rho + \beta\, \partial_u\left(\rho(1 - \rho)\right), & u > 0, \\ \rho(t, 0) = \dfrac{1}{2}. \end{cases} \tag{2.30}$$

Then, $\psi(t, \cdot)$ determined from $\rho(t, \cdot)$ by $\psi'(t, u) = -\rho(t, u)$ and $\psi(t, \infty) = 0$ belongs to the class X_R and satisfies the PDE (2.26).

4. *Cole-Hopf transformation at macroscopic level.* We first recall the well-known Cole-Hopf transformation. The solution $\rho(t, \cdot)$ of (2.30) is transformed under

$$\omega(t, u) := \exp\left\{\beta \int_u^\infty \rho(t, v)dv\right\}, \quad u \in \mathbb{R}_+, \tag{2.31}$$

into the solution of the linear heat equation

$$\begin{cases} \partial_t \omega = \partial_u^2 \omega + \beta \, \partial_u \omega, & u > 0, \\ 2\partial_u \omega(t, 0) + \beta \omega(t, 0) = 0, \\ \omega(t, \infty) = 1, \\ \omega(0, u) = \exp\left\{\beta \int_u^\infty \rho_0(v)dv\right\}. \end{cases} \tag{2.32}$$

In fact, simple computations show that

$$\partial_t \omega = \omega \beta \int_u^\infty \partial_t \rho dv = \omega \beta \left(-\partial_u \rho - \beta \rho(1 - \rho)\right),$$

$$\partial_u \omega = -\omega \beta \rho,$$

$$\partial_u^2 \omega = -\partial_u \omega \cdot \beta \rho - \omega \beta \partial_u \rho = \omega \beta (\beta \rho^2 - \partial_u \rho).$$

Therefore, the right-hand side of the first equation in (2.32) is given by

$$\omega \beta (\beta \rho^2 - \partial_u \rho - \beta \rho) = \partial_t \omega.$$

The boundary condition at $u = 0$ in (2.32) follows from $\rho(t, 0) = \frac{1}{2}$ and

$$\partial_u \omega(t, 0) = -\omega(t, 0)\beta \rho(t, 0) = -\omega(t, 0)\frac{\beta}{2}.$$

5. *Cole-Hopf transformation at microscopic level.* Let us consider the transformation at microscopic level:

$$\zeta_t(x) := \exp\left\{-(\log \varepsilon) \sum_{y=x}^\infty \eta_t(y)\right\}, \quad x \in \mathbb{N}.$$

Note that this corresponds to the macroscopic transformation $\rho \mapsto \omega$, since $-(\log \varepsilon) = -(\log \varepsilon_R(N)) \sim \frac{\beta}{N}$. Then, we have the following lemma:

Lemma 2.2. *The process $\zeta_t = (\zeta_t(x))_{x\in\mathbb{N}}$ satisfies the following stochastic equations:*

$$d\zeta_t(x) = N^2\big(\varepsilon\zeta_t(x-1) - (\varepsilon+1)\zeta_t(x) + \zeta_t(x+1)\big)dt + dM_t(x), \quad x \in \mathbb{N},$$

where $\zeta_t(0)$ is given by $\zeta_t(0) = \varepsilon^{-1}\zeta_t(2)$. Here, $M_t = (M_t(x))_{x\in\mathbb{N}}$ are martingales with quadratic variations given as follows:

$$\begin{cases} \dfrac{d}{dt}\langle M(x)\rangle_t = \zeta_t(x)^2\big\{a_N 1_{\{\eta_t(x-1)=1,\eta_t(x)=0\}} + b_N 1_{\{\eta_t(x-1)=0,\eta_t(x)=1\}}\big\}, & x \geq 2, \\[2mm] \dfrac{d}{dt}\langle M(1)\rangle_t = \zeta_t(1)^2\big\{a_N 1_{\{\eta_t(1)=0\}} + b_N 1_{\{\eta_t(1)=1\}}\big\}, \\[2mm] \langle M(x), M(y)\rangle_t = 0, \quad \text{if } 1 \leq x \neq y, \end{cases}$$

$$(2.33)$$

where

$$a_N = \frac{N^2}{\varepsilon}(1-\varepsilon)^2, \quad b_N = N^2(1-\varepsilon)^2,$$

both of which tend to β^2 as $N \to \infty$.

Proof. In general, if X_t is a process with a generator L, then for functions f and g defined on the state space of X_t, $M_t^f := f(X_t) - \int_0^t Lf(X_s)ds$ are martingales (with respect to a natural filtration) with quadratic variations described by the so-called carré du champ operator:

$$\frac{d}{dt}\langle M^f, M^g\rangle_t = 2\Gamma(f, g)(X_t) \equiv \{L(fg) - fLg - gLf\}(X_t).$$

The lemma is shown by applying this general principle. □

6. *Hydrodynamic limit for the Cole-Hopf transformed process.*

Proposition 2.2. *Set $\pi_t^N(du) = \frac{1}{N}\sum_{x\in\mathbb{N}}\zeta_t(x)\delta_{\frac{x}{N}}(du)$. Then, as $N \to \infty$, $\pi_t^N(du)$ converges to $w(t, u)du$ in probability, where $w(t, u)$ is the solution of (2.32).*

For the proof of this proposition, we need to show the tightness of $\{\pi_t^N\}_N$ and the uniqueness of the weak solution of (2.32) with the boundary condition. We skip these steps and explain only how the limit equation and the boundary condition can be derived.

(a) *Derivation of the limit equation.* Take any test function $f \in C_0^\infty(\mathbb{R}_+^\circ)$ and compute the stochastic derivative $d\langle\pi_t^N, f\rangle$. Then, by Lemma 2.2, its martingale term is given by $M_t^N = \frac{1}{N}\sum_{x\in\mathbb{N}}M_t(x)f(x/N)$ and satisfies

$$E[(M_t^N)^2] = \frac{1}{N^2}\sum_{x,y\in\mathbb{N}}E[M_t(x)M_t(y)]f(x/N)f(y/N) = O(1/N)$$

as $N \to \infty$. This asymptotic estimate can be shown by applying Lemma 2.2 and noting that $0 \leq \zeta_t(x) \leq \zeta_t(1) \sim e^{\beta X_t^N}$, where $X_t^N := \frac{1}{N}\sum_{y=1}^{\infty} \eta_t(y) = \frac{1}{N} \times$ the total number of particles, and X_t^N stays almost bounded as we will see later in (b). Lemma 2.2 also gives the drift term, which can be computed as follows:

$$\frac{1}{N}\sum_{x \in \mathbb{N}} N^2 \big(\varepsilon\zeta_t(x-1) - (\varepsilon + 1)\zeta_t(x) + \zeta_t(x+1)\big) f(x/N)$$

$$= \frac{1}{N}\sum_{x \in \mathbb{N}} N^2 \Big(\varepsilon f((x+1)/N) - (\varepsilon + 1)f(x/N) + f((x-1)/N)\Big)\zeta_t(x)$$

$$= \frac{1}{N}\sum_{x \in \mathbb{N}} \Big(-\beta f'(x/N) + f''(x/N) + O(1/N)\Big)\zeta_t(x)$$

$$= \langle -\beta f' + f'', \pi_t^N \rangle + o(1).$$

This leads to the weak form of the linear heat equation in the limit. The second equality follows by noting that

$$f((x \pm 1)/N) = f(x/N) \pm \frac{1}{N}f'(x/N) + \frac{1}{2N^2}f''(x/N) + O(1/N^3).$$

(b) *Derivation of the boundary condition.* The macroscopic boundary condition at 0 follows from the microscopically pointwise ergodicity:

Lemma 2.3. *For every $\delta, T > 0$, we have*

$$\lim_{N \to \infty} P\left(\left|\frac{1}{T}\int_0^T \eta_{N^2 s}(1)ds - \frac{1}{2}\right| > \delta\right) = 0.$$

It is usually difficult to show the ergodicity at a single point. Eyink, Lebowitz, and Spohn [76] proved the ergodicity under the spatial average near the boundary. The above lemma tells that $\eta_t(1) \in \{0, 1\}$ takes each value almost half and half over a microscopically long-time interval $[0, N^2 T]$ so that, under the time average, it converges to $\frac{1}{2}$. For the proof of Lemma 2.3, consider $X_t^N = \frac{1}{N}\{$total number of particles$\}$ introduced above. Then, its martingale part is given by

$$m_t^N = X_t^N - X_0^N - \int_0^t N^2 L_{\varepsilon R(N), R}(X_s^N)ds.$$

We can compute $N^2 L_{\varepsilon R(N), R}(X_s^N) = N(1 - 2\eta_s(1)) - \beta(1 - \eta_s(1))$, which is written only in $\eta_s(1)$, since the change of X_s^N occurs only at the boundary $x = 1$. This yields

$$\left|\int_0^T (1 - 2\eta_s(1))ds\right| \leq \frac{1}{N}(X_T^N + X_0^N + |m_T^N| + CT)$$

with some $C > 0$. However, the following estimates show that we can treat X_T^N and $|m_T^N|$ almost bounded:

$$\lim_{\lambda \to \infty} \sup_N P\left(\sup_{0 \le t \le T} X_t^N > \lambda \right) = 0 \quad \text{and} \quad E[(m_T^N)^2] \le T.$$

This completes the proof of Lemma 2.3.

2.2.3 Non-Equilibrium Fluctuations

We now consider the dynamical fluctuations of $\Psi_U^N(t, u) = \tilde{\psi}_{p_t}(u)$ and $\Psi_R^N(t, u) = \tilde{\psi}_{q_t}(u)$ around their hydrodynamic limits $\psi_U(t, u)$ and $\psi_R(t, u)$, respectively:

$$\Psi_U^N(t, u) := \sqrt{N}\big(\tilde{\psi}_U^N(t, u) - \psi_U(t, u)\big),$$
$$\Psi_R^N(t, u) := \sqrt{N}\big(\tilde{\psi}_R^N(t, u) - \psi_R(t, u)\big).$$

In general, the proof of non-equilibrium fluctuations is rather involved and only the result of Chang and Yau [48] for the Ginzburg-Landau model is known. But, in our case, due to a special feature of the model, we can prove the non-equilibrium fluctuations. The results of this section are due to Funaki, Sasada, Sauer, and Xie [115].

Theorem 2.8 (U-case). $\Psi_U^N(t, u)$ *converges weakly to* $\Psi_U(t, u)$ *as* $N \to \infty$ *on the space* $D([0, T], D(\mathbb{R}_+^\circ))$ *for every* $T > 0$. *The limit* $\Psi_U(t, u)$ *is a solution of the following SPDE:*

$$\partial_t \Psi(t, u) = \left(\frac{\Psi'(t, u)}{(1 + \rho_U(t, u))^2} \right)' + \alpha \frac{\Psi'(t, u)}{(1 + \rho_U(t, u))^2} + \sqrt{\frac{2\rho_U(t, u)}{1 + \rho_U(t, u)}} \dot{W}(t, u),$$

where $\rho_U(t, u) = -\psi_U'(t, u)$ *and* $\dot{W}(t, u)$ *is the space-time Gaussian white noise on* $[0, \infty) \times \mathbb{R}_+$, *see* (3.2).

Theorem 2.9 (RU-case). $\Psi_R^N(t, u)$ *converges weakly to* $\Psi_R(t, u)$ *as* $N \to \infty$ *on the space* $D([0, T], D(\mathbb{R}_+))$ *for every* $T > 0$. *The limit* $\Psi_R(t, u)$ *is a solution of the following SPDE:*

$$\partial_t \Psi(t, u) = \Psi''(t, u) + \beta(1 - 2\rho_R(t, u))\Psi'(t, u)$$
$$+ \sqrt{2\rho_R(t, u)(1 - \rho_R(t, u))} \dot{W}(t, u),$$
$$\Psi'(t, 0+) = 0,$$

where $\rho_R(t, u) = -\psi_R'(t, u)$.

We don't touch upon the proofs of these two theorems, but comment on the invariant measures of these SPDEs and clarify the relations with the static CLTs obtained in Sect. 2.1.3.

First in the U-case, since $\rho_U(t, u)$ converges to $\rho_U(u) := -\psi'_U(u)$ as $t \to \infty$, the SPDE in equilibrium has the form

$$\partial_t \Psi = -g_U(u) \, Q_U \Psi + \sqrt{2 g_U(u)} \, \dot{W}(t, u),$$

where

$$g_U(u) = \frac{\rho_U(u)}{1 + \rho_U(u)},$$

$$Q_U = -\frac{\partial}{\partial u}\left(\frac{1}{\rho_U(u)(1 + \rho_U(u))} \frac{\partial}{\partial u}\right) \quad \text{on} \quad L^2(\mathbb{R}^\circ_+, du).$$

Thus, the invariant measure of $\Psi_U(t, u)$ is $N(0, Q_U^{-1})$, the centered Gaussian measure with the covariance operator Q_U^{-1}, cf. the fact mentioned at the beginning of Sect. 5.3.2. However, one can easily check that $Q_U C_U(\cdot, v) = \delta_v(\cdot)$, which implies that $C_U(u, v)$ is the Green kernel of Q_U^{-1}. This gives an alternative proof of the static result, Theorem 2.2, in the U-case.

Secondly in the RU-case, since $\rho_R(t, u)$ converges to $\rho_R(u) := -\psi'_R(u)$ as $t \to \infty$, the SPDE in equilibrium reads

$$\partial_t \Psi = -g_R(u) \, Q_R \Psi + \sqrt{2 g_R(u)} \, \dot{W}(t, u),$$

where

$$g_R(u) = \rho_R(u)(1 - \rho_R(u)),$$

$$Q_R = -\frac{\partial}{\partial u}\left(\frac{1}{\rho_R(u)(1 - \rho_R(u))} \frac{\partial}{\partial u}\right) \quad \text{on} \quad L^2(\mathbb{R}_+, du),$$

with Neumann condition at $u = 0$. Thus, the invariant measure of $\Psi_R(t, u)$ is $N(0, Q_R^{-1})$. Since $C_R(u, v)$ is the Green kernel of Q_R^{-1}, which is shown by checking that $Q_R C_R(\cdot, v) = \delta_v(\cdot)$ and the Neumann condition at $u = 0$, this gives another proof of the static result, Theorem 2.2, in the RU-case.

2.2.4 Dynamic Large Deviation Principle

At the dynamic level, the following LDP is expected, cf. Kipnis, Olla, and Varadhan [169]. We give only a rough sketch of the proof. This section is based on a discussion with Bin Xie.

Theorem 2.10. *The LDP holds for* $\tilde{\psi}_U^N(t,u)$ *and* $\tilde{\psi}_R^N(t,u), t \in [0,T]$ *with speed N and rate functions*

$$I_U(\psi) = \frac{1}{4}\int_0^T dt \int_{\mathbb{R}_+^\circ} \left\{\partial_t \psi(t,u) - \left(\frac{\psi'(t,u)}{1-\psi'(t,u)}\right)' - \alpha\frac{\psi'(t,u)}{1-\psi'(t,u)}\right\}^2 \frac{1-\psi'(t,u)}{-\psi'(t,u)} du,$$

and

$$I_R(\psi) = \frac{1}{4}\int_0^T dt \int_{\mathbb{R}_+} \frac{\{\partial_t \psi(t,u) - \psi''(t,u) - \beta\psi'(t,u)\big(1+\psi'(t,u)\big)\}^2}{-\psi'(t,u)\big(1+\psi'(t,u)\big)} du,$$

respectively, for $\psi = \psi(t,u)$ *satisfying* $\psi'(t,u) \le 0$ (*otherwise* $I_U(\psi), I_R(\psi) = \infty$).

Rough sketch of the proof. We first consider the RU-case. From work of Bertini, Landim, and Mourragui [28], one can expect that the LDP holds for the height differences and the rate function for $\rho(t,u)\big(= -\psi'(t,u)\big)$ is given by

$$I(\rho) = \frac{1}{4}\|\partial_t \rho - \rho'' - \beta\big(\rho(1-\rho)\big)'\|_{-1,\rho}^2,$$

where

$$\|L\|_{-1,\rho}^2 = \sup_{G\in C_0^\infty((0,T)\times\mathbb{R}_+^\circ)} \left\{2\langle L,G\rangle - \|G\|_{1,\rho}^2\right\},$$

$$\langle L,G\rangle = \int_0^T dt \int_{\mathbb{R}_+} L(t,u)G(t,u)du,$$

$$\|G\|_{1,\rho}^2 = \int_0^T dt \int_{\mathbb{R}_+} \big(G'(t,u)\big)^2 \rho(t,u)\big(1-\rho(t,u)\big)du.$$

Indeed, [28] treated the case of a bounded interval with stochastic reservoirs at the two endpoints, but we can expect a similar result in our situation too; note that the LDP is discussed on an unbounded domain by Landim and Yau [174]. In [28], the jump rates to the right and the left behave as $1 + \frac{E}{2N}$ and $1 - \frac{E}{2N}$ as $N \to \infty$, respectively, while these are $1 - \frac{\beta}{N}$ and 1 in our case. This implies that we can take $E = -\beta$ asymptotically. Moreover, the generator of the WASEP considered in [28] has the front factor $\frac{1}{2}$ compared with ours. This requires a proper modification on the rate function $I(\rho)$.

We define $I(\psi) := I(\rho)$ with $\rho(t,u) = -\psi'(t,u)$. Rewriting $I(\rho)$ in terms of ψ and applying integration by parts, we obtain

$$I(\psi) = \frac{1}{4} \sup_G \left[2 \int_0^T dt \int_{\mathbb{R}_+} \{\partial_t \psi(t,u) - \psi''(t,u) - \beta \psi'(t,u)(1 + \psi'(t,u))\} \right.$$

$$\left. \times (-G'(t,u))du - \int_0^T dt \int_{\mathbb{R}_+} (-\psi'(t,u))(1 + \psi'(t,u))(G'(t,u))^2 du \right].$$

Setting $F(t,u) = -G'(t,u)\sqrt{-\psi'(t,u)(1 + \psi'(t,u))}$, we have that

$$I(\psi) = \frac{1}{4} \sup_F \left[2 \int_0^T dt \int_{\mathbb{R}_+} \frac{\partial_t \psi(t,u) - \psi''(t,u) - \beta \psi'(t,u)(1 + \psi'(t,u))}{\sqrt{-\psi'(t,u)(1 + \psi'(t,u))}} du \right.$$

$$\left. - \int_0^T dt \int_{\mathbb{R}_+} F(t,u)^2 du \right].$$

We easily see that $I(\psi) = I_R(\psi)$ by completing square, and this concludes the proof for the RU-case.

For the U-case, we assume that the assertion for the RU-case holds for the height differences defined on the whole lattice $\mathbb{Z}$ (even though this is an unbounded domain) and the rate function is similarly given; we should replace $\mathbb{R}_+$ by $\mathbb{R}$ and β by α, respectively. Then, applying the nonlinear transformation $\rho = \Phi_U(\psi) \in Y_U$ for $\rho \in X_U$ defined by Lemma 2.1, we get the conclusion.   □

Let us discuss the relations between static and dynamic rate functions, cf. Proposition 8.1 of Funaki and Nishikawa [109], Bertini, De Sole, Gabrielli, Jona Lasinio, and Landim [26]. For $T > 0$ and $\bar{\psi} = \{\bar{\psi}(u); u \in \mathbb{R}_+^\circ\}$, set

$$S_U^T(\bar{\psi}) := \inf_{\psi = \psi(t,u) \text{ s.t. } \psi(T,u) = \bar{\psi}(u)} I_U^T(\psi),$$

where I_U^T is the dynamic rate function up to time T given by Theorem 2.10. By the contraction principle, S_U^T is the large deviation rate function for the distribution of the scaled height functions $\tilde{\psi}_U^N(T,u)$ at time T. We similarly define $S_R^T(\bar{\psi})$ in the RU-case. Then, we can recover the static rate functions $\mathbb{F}_U$ and $\mathbb{F}_R$ obtained in Theorem 2.4 from the dynamic rate functions:

Proposition 2.3. *We have*

$$\lim_{T \to \infty} S_U^T(\bar{\psi}) = \mathbb{F}_U(\bar{\psi}) \quad and \quad \lim_{T \to \infty} S_R^T(\bar{\psi}) = \mathbb{F}_R(\bar{\psi}).$$

Proof. We first consider the U-case. The infimum of $I_U^T(\psi)$ is attained by the time-reversed classical trajectory. Indeed, let $\hat{\psi}(t,u), 0 \le t \le T$, be the solution of the hydrodynamic equation (2.23) with initial value $\hat{\psi}(0, u) = \bar{\psi}(u)$, and consider its time-reversal $\psi(t, u) = \hat{\psi}(T - t, u), 0 \le t \le T$. Then we have

$$\frac{d}{dt}\mathbb{F}_U(\psi(t)) = \int_{\mathbb{R}_+^\circ} \frac{\delta\mathbb{F}_U}{\delta\psi(u)}(\psi(t)) \, \partial_t\psi(t, u)du, \quad 0 \le t \le T,$$

and, since we can easily verify that

$$\frac{\delta\mathbb{F}_U}{\delta\psi(u)} = h_U''(-\psi'(u))\psi''(u) + \alpha = \frac{\psi''(u)}{\psi'(u)\big(1 - \psi'(u)\big)} + \alpha,$$

by the hydrodynamic equation (2.23), we obtain

$$\frac{\delta\mathbb{F}_U}{\delta\psi(u)}(\psi(t)) = \frac{1 - \psi'(t, u)}{\psi'(t, u)} \left\{ \frac{\psi''(t, u)}{\big(1 - \psi'(t, u)\big)^2} + \alpha \frac{\psi'(t, u)}{1 - \psi'(t, u)} \right\}$$

$$= \frac{1 - \psi'(t, u)}{-\psi'(t, u)} \partial_t\psi(t, u).$$

Therefore, we have that

$$\frac{d}{dt}\mathbb{F}_U(\psi(t)) = \int_{\mathbb{R}_+^\circ} \big(\partial_t\psi(t, u)\big)^2 \frac{1 - \psi'(t, u)}{-\psi'(t, u)} du$$

$$= \int_{\mathbb{R}_+^\circ} \left\{ \frac{\psi''(t, u)}{\big(1 - \psi'(t, u)\big)^2} + \alpha \frac{\psi'(t, u)}{1 - \psi'(t, u)} \right\}^2 \frac{1 - \psi'(t, u)}{-\psi'(t, u)} du$$

$$= -\int_{\mathbb{R}_+^\circ} \partial_t\psi(t, u) \left\{ \frac{\psi''(t, u)}{\big(1 - \psi'(t, u)\big)^2} + \alpha \frac{\psi'(t, u)}{1 - \psi'(t, u)} \right\} \frac{1 - \psi'(t, u)}{-\psi'(t, u)} du.$$

Thus, by the definition of $I_U^T(\psi)$,

$$\mathbb{F}_U(\bar{\psi}) - \mathbb{F}_U(\hat{\psi}(T)) = I_U^T(\psi).$$

The proof for the U-case is completed by noting that $\lim_{T\to\infty} \mathbb{F}_U(\hat{\psi}(T)) = 0$.

The proof for the RU-case is similar. In the RU-case, since we can easily check that

$$\frac{\delta\mathbb{F}_R}{\delta\psi(u)} = h_R''(-\psi'(u))\psi''(u) + \beta = \frac{\psi''(u)}{\psi'(u)\big(1 + \psi'(u)\big)} + \beta,$$

by the hydrodynamic equation (2.26), we have that

$$\frac{\delta \mathbb{F}_R}{\delta \psi(u)}(\psi(t)) = \frac{1}{\psi'(t,u)\big(1 + \psi'(t,u)\big)} \big\{ \psi''(t,u) + \beta \psi'(t,u)\big(1 + \psi'(t,u)\big)\big\}$$

$$= \frac{1}{-\psi'(t,u)\big(1 + \psi'(t,u)\big)} \partial_t \psi(t,u).$$

This shows that

$$\mathbb{F}_R(\bar{\psi}) - \mathbb{F}_U(\hat{\psi}(T)) = I_R^T(\psi)$$

for the time-reversed trajectory of the solution of the hydrodynamic equation. This completes the proof for the RU-case, since $\lim_{T \to \infty} \mathbb{F}_R(\hat{\psi}(T)) = 0$.   □

2.2.5 Conservative Systems

Up to now we discussed the dynamics associated with the U- and RU-grand canonical ensembles, so that the area of the Young diagrams is not conserved by the time evolution. Here we introduce a dynamics associated with the canonical ensemble, in particular, the RU-canonical ensemble. This dynamics conserves the area of the Young diagrams. It is related to the model called *surface diffusion* [193] in the sense that the unit squares (considered as atoms) move sliding on the surface (interface), see Fig. 2.15.

Let $\mathcal{Q}_{K,M}$ be the set of all sequences (distinct partitions) $q = \{q_1 > q_2 > \cdots > q_K \geq 1\} \in \mathcal{Q}_M$ of $M \in \mathbb{N}$, which is a variant of $\mathcal{Q}_M, \mathcal{Q}_{K,M,L}$ considered in Sects. 2.1.1 and 2.1.5. Recall that, by (2.1), a sequence (partition) $q \in \mathcal{Q}_{K,M}$ is identified with the height function ψ_q of a 2D Young diagram satisfying that $\psi_q(0) = K$ and $\int_0^\infty \psi_q(u)du = M$. We denote by $\mu_{K,M}$ the uniform probability measure on $\mathcal{Q}_{K,M}$.

Dynamics: We construct a random dynamics $q(t)$ on $\mathcal{Q}_{K,M}$, which is reversible under $\mu_{K,M}$, in such a manner that the corresponding dynamics of the Young diagrams $\psi_{q(t)}$ has the following properties: The dynamics conserves the area of the Young diagrams, i.e., the creation and annihilation of unit squares take place simultaneously. (The dynamics we have discussed up to now does not enjoy this

Fig. 2.15 Conservative
dynamics – surface diffusion

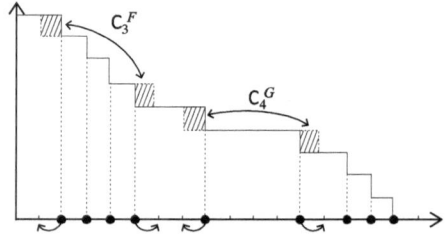

property and creation and annihilation occur independently.) Or, a unit square moves on the surface of the Young diagram until it finds another stable position keeping height differences $\in \{0,1\}$. The jump rates of a square falling down a stair with length r and its reversed transition are $c_r^F > 0$. The jump rates of a square sliding over a flat piece of length r and its reversed transition are $c_r^G > 0$. In particular, the transitions are local and this character of the dynamics is quite natural from the physical point of view.

To describe such dynamics, we adopt a particle picture in view of the hight differences introduced in (2.9) and (2.20). Let $\mathscr{X}_{K,M}$ be the set of all $\eta = \{\eta_k\}_{k \in \mathbb{N}} \in \mathscr{X} := \{0,1\}^{\mathbb{N}}$ satisfying the two conditions in (2.10) and let $\nu_{K,M}$ be the uniform probability measure on $\mathscr{X}_{K,M}$. Recall that one can identify the two spaces $\mathscr{Q}_{K,M}$ and $\mathscr{X}_{K,M}$ by means of the map $\psi_q \in \mathscr{Q}_{K,M} \mapsto \eta \in \mathscr{X}_{K,M}$ determined by (2.9), and $\nu_{K,M}$ is the image measure of $\mu_{K,M}$ under this map.

For $\eta \in \mathscr{X}$ and $i,j \in \mathbb{N}$ such that $i \neq j$, we define a new configuration $\eta^{i,j} \in \mathscr{X}$ by interchanging the states at the sites i and j, that is,

$$(\eta^{i,j})_k = \begin{cases} \eta_j, & k = i, \\ \eta_i, & k = j, \\ \eta_j, & k \neq i,j, \end{cases}$$

and for $i,j \in \mathbb{N}$ such that $2 \leq i < j$, we define

$$\sigma^{(i,j)}\eta \equiv \sigma^{i-1,i;j,j+1}\eta := (\eta^{i-1,i})^{j,j+1},$$

the configuration obtained by interchanging the states at $i-1$ and i, and those at j and $j+1$ simultaneously. For functions $f : \mathscr{X} \to \mathbb{R}$, define

$$\pi^{(i,j)}f(\eta) \equiv \pi^{i-1,i;j,j+1}f(\eta) := f\big(\sigma^{(i,j)}\eta\big) - f(\eta).$$

For $r = 1, 2, \ldots$, define four functions $F_r^{\pm}, G_r^{\pm}$ on $\tilde{\mathscr{X}} := \{0,1\}^{\mathbb{Z}}$ by

$$F_r^+(\eta) = 1_{\{\eta_{-1}=0, \eta_0=\cdots=\eta_r=1, \eta_{r+1}=0\}},$$

$$F_r^-(\eta) \equiv F_r^+(\sigma^{(0,r)}\eta) = 1_{\{\eta_{-1}=1, \eta_0=0, \eta_1=\cdots=\eta_{r-1}=1, \eta_r=0, \eta_{r+1}=1\}},$$

$$G_r^+(\eta) \equiv F_r^-(\check{\eta}) = 1_{\{\eta_{-1}=0, \eta_0=1, \eta_1=\cdots=\eta_{r-1}=0, \eta_r=1, \eta_{r+1}=0\}},$$

$$G_r^-(\eta) \equiv F_r^+(\check{\eta}) = 1_{\{\eta_{-1}=1, \eta_0=\cdots=\eta_r=0, \eta_{r+1}=1\}},$$

where $\sigma^{(i,j)}\eta$ is defined similarly as above for $\eta \in \tilde{\mathscr{X}}$ and $i,j \in \mathbb{Z}$ such that $i < j$, and $\check{\eta} \in \tilde{\mathscr{X}}$ is the configuration determined from $\eta \in \tilde{\mathscr{X}}$ by $\check{\eta}_i = 1 - \eta_i$ for all $i \in \mathbb{Z}$. The shift operators $\tau_i : \tilde{\mathscr{X}} \to \tilde{\mathscr{X}}$ are defined by $(\tau_i\eta)_k := \eta_{k+i}$ for $\eta \in \tilde{\mathscr{X}}$, $i, k \in \mathbb{Z}$, and they induce operators acting on functions $F : \tilde{\mathscr{X}} \to \mathbb{R}$ as $\tau_i F(\eta) := F(\tau_i\eta)$.

Let non-negative bounded constants $(c_r^F)_{r \geq 1}$ and $(c_r^G)_{r \geq 1}$ be given and consider the operator L acting on functions $f : \mathscr{X} \to \mathbb{R}$ as

$$Lf(\eta) = \sum_{2 \le i < j} c_{(i,j)}(\eta) \pi^{(i,j)} f(\eta), \tag{2.34}$$

where

$$c_{(i,j)}(\eta) = c_{j-i}^F \tau_i (F_{j-i}^+ + F_{j-i}^-) + c_{j-i}^G \tau_i (G_{j-i}^+ + G_{j-i}^-).$$

We may assume $c_1^G = 0$, since $F_1^{\pm} = G_1^{\pm}$, so that the transitions under $F_1^{\pm}$ are the same as those under $G_1^{\pm}$. Note that, for $i, j \in \mathbb{N}$ such that $2 \le i < j$, $\tau_i F_{j-i}^{\pm}$ and $\tau_i G_{j-i}^{\pm}$ are functions on $\mathscr{X}$. For example, we have that

$$\tau_i F_{j-i}^+(\eta) = 1_{\{\eta_{i-1}=0, \eta_i = \cdots = \eta_j = 1, \eta_{j+1} = 0\}}.$$

The operator L determines a Markov process $\eta(t)$ on $\mathscr{X}$. This is a kind of exclusion process on $\mathbb{N}$, which has the special feature that the jumps are made simultaneously by two different particles to their neighboring sites in such a manner that, if one particle jumps to the right, the other moves to the left. More precisely, the time evolution of particles is governed by the following two rules:

1. two particles at the ends of consecutively lined $r + 1$ particles jump simultaneously to outward directions at rate c_r^F if two visiting sites are vacant, and the reversed transition occurs at the same rate c_r^F; and
2. two particles at the ends of consecutive $r + 1$ vacant sites jump simultaneously to inward directions at rate c_r^G, and the reversed transition occurs at the same rate c_r^G.

Under these transitions, the two quantities in (2.10) are always conserved, so that $\eta(0) \in \mathscr{X}_{K,M}$ implies $\eta(t) \in \mathscr{X}_{K,M}$ for all $t > 0$. Under the map $\mathscr{X}_{K,M} \ni \eta \mapsto \psi_q \in \mathscr{Q}_{K,M}$, the Markov process $\eta(t)$ defines an area-preserving random dynamics of Young diagrams $\psi_{q(t)}$ or $q(t)$ in $\mathscr{Q}_{K,M}$.

It is natural to assume that each atom located on the surface of the Young diagrams moves over the surface like a Brownian particle. Then, the jump rates c_r^F and c_r^G would behave as C/r^2 with $C > 0$ as r grows.

The operator L is reversible under $\nu_{K,M}$ on $\mathscr{X}_{K,M}$ and also under the Bernoulli measures ν_ρ on $\mathscr{X}$ for all $\rho \in [0, 1]$, which represents the mean of ν_ρ. In fact, for $\nu_{K,M}$, the corresponding Dirichlet form is given by

$$\mathscr{E}_{K,M}(f, g) := -E^{\nu_{K,M}}[Lf \cdot g] = \frac{1}{2} \sum_{2 \le i < j} E^{\nu_{K,M}}[c_{(i,j)} \pi^{(i,j)} f \, \pi^{(i,j)} g],$$

for $f, g : \mathscr{X}_{K,M} \to \mathbb{R}$. Note that the sum can be restricted to $2 \le i < j \le M - 1$, since $\eta \in \mathscr{X}_{K,M}$ implies $\eta_j = 0$ for all $j \ge M + 1$, so that $c_{(i,j)}(\eta) = 0$ if $j \ge M + 1$.

Hydrodynamic limits: One of the interesting problems to study is the hydrodynamic behavior of the Markov process $\eta(t)$ generated by L, or the corresponding time evolution $\psi_{q(t)}$ of the Young diagrams. Taking the spatial size $N \in \mathbb{N}$ of the system as a scaling parameter, the result of Theorem 2.5 suggests that K and M

should be scaled as $K = c_1 N$ and $M = c_2 N^2 \in \mathbb{N}$ with some fixed $c_1, c_2 > 0$. Furthermore, the scaling in time is expected to be N^4, so that we consider the process $\eta^N(t)$ on $\mathscr{X}_{c_1 N, c_2 N^2}$ with generator $L^N = N^4 L$.

More precisely, we consider the associated particle system on a torus: $\eta(k) \in \{0, 1\}$, $k \in \mathbb{T}_N = \mathbb{Z}/N\mathbb{Z}$, to avoid the complexity of working on an infinite region. Define the empirical measure ξ_t^N under the time change $t \mapsto N^4 t$ and spatial scale change $k \mapsto u = k/N$:

$$\xi_t^N(du) = \frac{1}{N} \sum_{k \in \mathbb{N}} \eta_{N^4 t}(k) \delta_{k/N}(du), \quad u \in \mathbb{T} = [0, 1].$$

Our conjecture is that $\xi_t^N(du)$ converges to $\rho(t, u)du$ as $N \to \infty$ and the limit density $\rho(t, u)$ satisfies the Cahn-Hilliard type fourth-order nonlinear PDE of parabolic type (cf. Chaps. 3 and 4):

$$\frac{\partial \rho}{\partial t} = -\frac{\partial^2}{\partial u^2} \left\{ D(\rho) \frac{\partial^2 \rho}{\partial u^2} \right\}, \quad u \in \mathbb{T},$$

where

$$D(\rho) = \frac{1}{\rho(1 - \rho)} \inf_{f:\text{tame}} \frac{1}{4} \sum_{r=1}^{\infty} \left\langle c_{(0,r)} \left\{ \pi^{(0,r)} \left(\Gamma_f + \frac{1}{2} \sum_k k^2 \eta(k) \right) \right\}^2 \right.$$
$$\left. + c_{(-r,0)} \left\{ \pi^{(-r,0)} \left(\Gamma_f + \frac{1}{2} \sum_k k^2 \eta(k) \right) \right\}^2 \right\rangle_\rho,$$

where $\Gamma_f = \sum_k \tau_k f$, $\langle \cdot \rangle_\rho$ denotes the average under ν_ρ, $c_{(0,\pm r)}$ are jump rates determined by c_r^F, c_r^G, and $\pi^{(0,\pm r)}$ are the transition operators explained above.

In a derivation of such an equation, we compute $L^N \eta_k$ as usual and a diverging factor N^4 appears due to the time change. In fact, the factor N^2 can be absorbed freely, but for an additional N^2, we need to replace a seemingly diverging term of order N^2 describing currents with a better converging term of order 1. In the theory of the hydrodynamic limit, in deriving second-order parabolic equations, such replacement is called the gradient replacement, see [168, 211]. However, in our setting, replacing the gradient is not enough, and instead its elaboration, the Laplacian replacement, is required. In fact, the Laplacian replacement (the fluctuation-dissipation relation) for the current W would have the form

$$W = -D(\rho)(\eta_1 - 2\eta_0 + \eta_{-1}) + Lf,$$

with some $f = f(\eta)$, where the current W is defined by $W = \sum_{r=1}^{\infty}(r + 1)W_r$ with $W_r = c_r^F(F_r^+ - F_r^-) + c_r^G(G_r^+ - G_r^-)$.

The equivalence of ensembles discussed in Sect. 2.1.5 is required as one of the steps in the proof of the hydrodynamic limit. Another important step is deriving the spectral gap of order $O(N^{-4})$ for our generator L defined on $\mathscr{X}_{c_1 N, c_2 N^2}$ under certain conditions on the jump rates c_r^F and c_r^G. This is recently proved by Nagahata [190].

Conservative dynamics of 2D Young diagrams of mean field type were studied by Baccelli, Karpelevich, Kelbert, Puhalskii, Rybko, and Suhov [16], Rybko, Shlosman, and Vladimirov [197], and Hora [147]. The transition rule in their dynamics is nonlocal and different from ours.

2.3 3D Young Diagrams

2.3.1 Static Theory

We call $p = \{p_{ij}\}_{1 \le i,j < \infty}$ a planar partition of $n \in \mathbb{N}$ if $p_{ij} \in \mathbb{Z}_+$, $\sum_{i,j} p_{ij} = n$, and $p_{ij} \ge p_{i+1,j}$ and $p_{ij} \ge p_{i,j+1}$ hold for all i,j. We denote the set of all such p's by $\mathscr{P}_n$. Once $p \in \mathscr{P}_n$ is given, one can associate a three-dimensional Young diagram ψ_p with volume n by pilling up p_{ij} unit boxes on the unit square with center $(i - \frac{1}{2}, j - \frac{1}{2})$, as indicated as in Fig. 2.16.

For fixed $N \in \mathbb{N}$, let Y_N be a random Young diagram with volume N^3. Namely, we consider a uniform probability on $\mathscr{P}_{N^3}$ and define $Y_N = \psi_p$ with random p. In other words, we consider the canonical ensemble of the uniform statistics. The limit shape of scaled surfaces of 3D Young diagrams under such ensemble was studied by Cerf and Kenyon [45].

Theorem 2.11 (LLN). *The scaled surface $\frac{1}{N}\{surface \ of \ Y_N\}$ of the random Young diagram converges to $\left(4/\zeta(3)\right)^{1/2} S_0$ as $N \to \infty$, where $\zeta(3) = \sum_{n=1}^{\infty} \frac{1}{n^3}$. The surface S_0 is in the first quadrant $(\mathbb{R}_+)^3$ of $\mathbb{R}^3$ and defined as follows:*

$$S_0 = \{(f(A, B, C) - \log A, f(A, B, C) - \log B, f(A, B, C) - \log C) \in \mathbb{R}^3; \ A, B, C > 0\},$$

where $f(A, B, C) = \frac{1}{4\pi^2} \int_0^{2\pi} \int_0^{2\pi} \log |A + Be^{iu} + Ce^{iv}| du dv$.

Fig. 2.16 3D Young diagram constructed from p

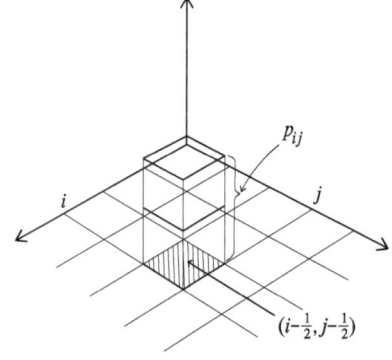

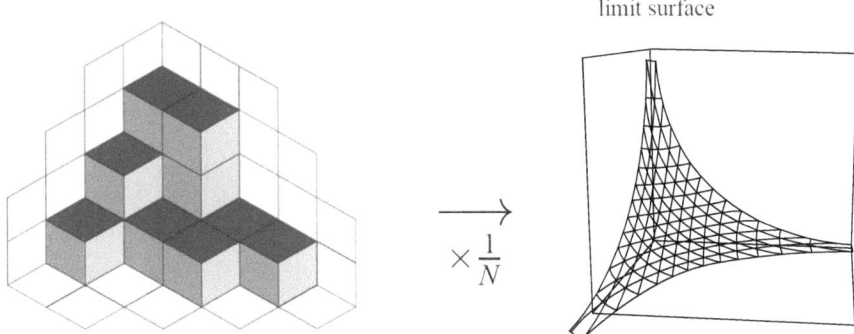

limit surface

Fig. 2.17 Scaling limit in 3D (Taken from [45]), see also Fig. 2.18

Cerf and Kenyon [45] also proved the corresponding LDP. If we replace (A, B, C) by (cA, cB, cC) for any $c > 0$, the formula defining S_0 given in this theorem does not change. This means that S_0 is parametrized essentially by two parameters, so that S_0 is a surface in $(\mathbb{R}_+)^3$. The surface S_0 touches the xy-plane (and also the yz-, zx-planes) and, after it touches, S_0 stays on these planes, see the right drawing in Fig. 2.17. Such a border is called a facet edge or the facet border. S_0 is a C^1-surface, but at the place it touches these planes (i.e., at the facet edge), it has singularities and is not a C^2-surface. The three curves appearing at the facet edge are actually the Vershik curve. In fact, $S_0 \cap \{xy\text{-plane}\} = \{(\log \frac{C}{A}, \log \frac{C}{B}, 0)\}$, since $f(A, B, C) = \log C$ when the third coordinate of S_0 is 0. One can see that this relation is equivalent to $A + B \leq C$, and setting $x = \log \frac{C}{A}$, $y = \log \frac{C}{B}$, we see that $S_0 \cap \{xy\text{-plane}\} = \{(x, y); e^{-x} + e^{-y} = \frac{A}{C} + \frac{B}{C} \leq 1\}$. The boundary of this planar region is described by $e^{-x} + e^{-y} = 1$, which is exactly the Vershik curve.

The surface S_0 has a variational characterization (isoperimetric problem) and is considered as a variant of the Wulff shape (crystal). Let $\mathscr{E}$ denote the family of all closed sets A of $(\mathbb{R}_+)^3$ having finite volume $|A| < \infty$ and satisfying the monotonicity property: if $(x, y, z) \in A$, then $[0, x] \times \{y\} \times \{z\} \subset A$, $\{x\} \times [0, y] \times \{z\} \subset A$, and $\{x\} \times \{y\} \times [0, z] \subset A$. Let $\mathscr{E}_1 = \{A \in \mathscr{E}; |A| \leq 1\}$. For $A \in \mathscr{E}$ with a smooth boundary $\partial_+ A(:= \partial A \cap (\mathbb{R}_+^\circ)^3)$ in the first quadrant, we define its surface energy $\Sigma(A)$ (or the total surface tension, or entropy, as called by [45]) by

$$\Sigma(A) = \int_{\partial_+ A} \sigma(\nu_A(x)) d\mathscr{H}^2(x),$$

where $\nu_A(x) \in S^2$ is the outward normal vector at $x \in \partial A$, $\mathscr{H}^2$ is the two-dimensional Hausdorff measure on $\partial_+ A$ and $\sigma(\nu)$ is the direction-dependent *surface tension* defined as follows:

$$\sigma(\nu) = -|\nu|_1 \text{ent}(\nu), \quad \nu = (a, b, c) \in S^2,$$

$$|v|_1 = |a| + |b| + |c|,$$

$$\text{ent}(v) = \frac{1}{\pi} L\left(\pi \frac{|a|}{|v|_1}\right) + L\left(\pi \frac{|b|}{|v|_1}\right) + L\left(\pi \frac{|c|}{|v|_1}\right),$$

$$L(x) = -\int_0^x \log(2\sin t)dt, \quad x \in [0, 2\pi),$$

is the Lobachevsky function. Then, S_0 provides the minimum of the following variational problem:

$$\inf_{A \in \mathscr{E}_1, \partial_+ A:\text{smooth}} \Sigma(A).$$

In Theorem 2.11, we have discussed the case of the canonical ensemble, but we can also consider the case of the grand canonical ensemble. Namely, let $\mathscr{P} = \bigcup_{n=0}^{\infty} \mathscr{P}_n$ and define a probability measure $\mu_{\mathsf{U}}^{\varepsilon}$, $0 < \varepsilon < 1$ on $\mathscr{P}$ by $\mu_{\mathsf{U}}^{\varepsilon}(p) = \frac{1}{Z(\varepsilon)} \varepsilon^{n(p)}$, $p \in \mathscr{P}$, as in (2.2) for the 2D case, where $n(p) = n$ for $p \in \mathscr{P}_n$ (i.e., $n(p)$ is the volume of the corresponding 3D Young diagram ψ_p) and $Z(\varepsilon)$ is the normalization constant.

Lemma 2.4. *Define* $\varepsilon = \varepsilon_{\mathsf{U}}(N)$ *by* $E^{\mu_{\mathsf{U}}^{\varepsilon}}[n(p)] = N^3$. *Then, we have* $\varepsilon_{\mathsf{U}}(N) = 1 - \frac{\alpha}{N} + \cdots$, *with* $\alpha = \frac{2}{3} \max\{\text{ent}(A); A \in \mathscr{E}_1\}$, *where* $\text{ent}(A) = \int_{\text{dom}(A)} \text{ent}(v_A(x))d\lambda(x)$, *see* [45], p. 157 *for more details.*

Indeed, [45] showed the asymptotic behavior $\sharp \mathscr{P}_n \sim e^{Cn^{2/3}}$ as $n \to \infty$, with $C = \max\{\text{ent}(A); A \in \mathscr{E}_1\}$. This is a three-dimensional version of the Hardy-Ramanujan formula. From this, similarly to the 2D case, the Laplace method shows that $\alpha = \frac{2}{3}C$ and the lemma follows.

The fluctuation of the facet edge was studied by Ferrari and Spohn [80]. It is shown that the order of the fluctuation is $O(N^{1/3})$ and the scaling limit is described by the Airy process. Figure 2.18 shows the facets whose interior is the first quadrant of $\mathbb{R}^3$ from which the solid figure shown in the right drawing in Fig. 2.17 is removed. Duits [66] shows that the fluctuation in the smooth part of the surface is given by the Gaussian free field (massless field) given by a different model.

Fig. 2.18 Crystal facets
(Taken from [80])

2.3.2   Dynamic Theory

As we saw, 2D Young diagrams in the U-case determine a particle system on $\mathbb{N}$ through the height differences. The Russian transformation projects this system onto the line $\{x+y = 0\}$ and recasts it as the exclusive particle system on $\mathbb{Z}$. In particular, a 2D Young diagram is essentially a 1D system.

For the 3D Young diagrams, three types of faces (horizontal, facing left, facing right) of unit boxes correspond to the height differences in the 2D case. In $\mathbb{R}^3$, one can project a Young diagram onto the plane $\{x + y + z = 0\}$ from the direction of $(1, 1, 1)$, see the left drawing in Fig. 2.17 (dark black: horizontal, light black: facing left, pale black: facing right). This corresponds to the Russian transformation. In this way, 3D Young diagrams can be transformed into lozenge (rhombic) tiling or dimer configurations (dimer covers) on a honeycomb lattice, see Fig. 2.19. Conversely, the height functions can be recovered from the dimer configurations, see Fig. 2.25 below. We now explain this terminology in more detail.

Honeycomb lattice: Let G_∞ be the infinite *honeycomb lattice* as is shown in the left drawing of Fig. 2.20 This is a bipartite planar graph, i.e., the vertices (sites) are divided into two classes, even and odd (● and ○ in the right drawing of Fig. 2.20), and the lattice is embedded in the plane. It is $\mathbb{Z}^2$-periodic. Namely, assuming the length of one side of each *hexagon* is 1, take the origin of the plane at the center

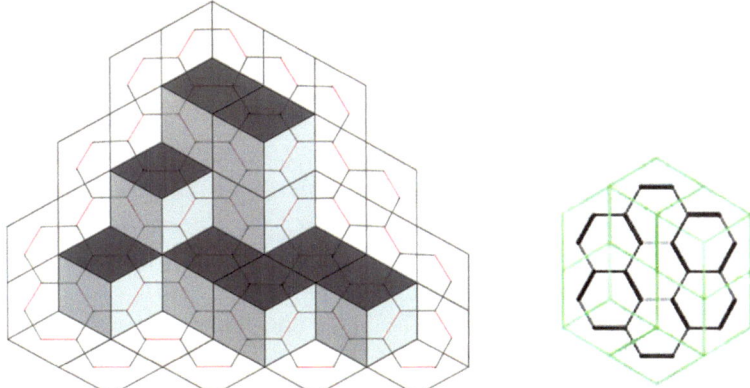

Fig. 2.19 3D Young diagram and corresponding dimer cover

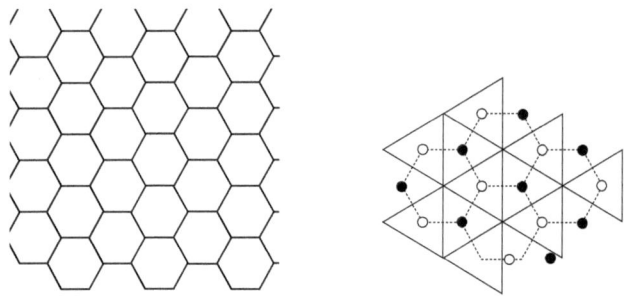

Fig. 2.20 Honeycomb lattice G_∞ and its dual lattice (triangular lattice, Taken from [45])

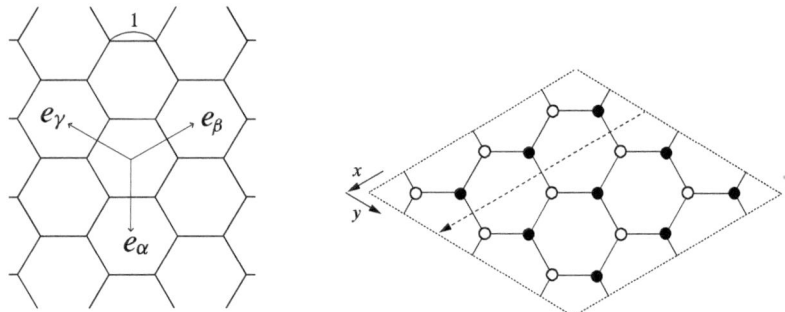

Fig. 2.21 $e_\alpha, e_\beta, e_\gamma$ **Fig. 2.22** H_3

of one hexagon and consider three vectors $e_\alpha = (0, -\sqrt{3}), e_\beta = (\frac{3}{2}, \frac{\sqrt{3}}{2}), e_\gamma = (-\frac{3}{2}, \frac{\sqrt{3}}{2}) \in \mathbb{R}^2$ as in Fig. 2.21. Then, G_∞ is invariant under the shift in the direction of $m e_\beta + n e_\gamma$ with $(m, n) \in \mathbb{Z}^2$.

We also consider the *honeycomb torus* $H_N = G_\infty / N\mathbb{Z}^2$ for $N \in \mathbb{N}$, see Fig. 2.22 for H_3. H_N consists of N^2 hexagons and is useful to consider the dynamics on it, since infinite systems are sometimes difficult to analyze. Let H_N^B be the family of all undirected edges (bonds) of H_N. Then, $\sharp H_N$ (the number of vertices of H_N) = $2N^2 (= 6 \times \frac{1}{3}N^2)$, since one hexagon has six vertices and one vertex is shared by three hexagons, while $\sharp H_N^B = 3N^2$, since one hexagon has six edges and one edge is shared by two hexagons. There are three different types α, β, γ of edges in H_N shown in Fig. 2.26. Indeed, we say that $b \in H_N$ is of α-type if it is horizontal (i.e., parallel to the x-axis or to the vector $(1, 0)$), β-type if it is parallel to the vector $(-\frac{1}{2}, \frac{\sqrt{3}}{2})$, and γ-type if it is parallel to the vector $(\frac{1}{2}, \frac{\sqrt{3}}{2})$.

Let H_N^* be the dual lattice of H_N, see the right drawing in Fig. 2.20. It is a *triangular lattice* and $i \in H_N^*$ represents the center of a hexagon of the original lattice H_N. Let $(H_N^*)^B$ be the family of all undirected edges of H_N^*. Then, there exists a one-to-one correspondence between edges $b \in H_N^B$ and $b' \in (H_N^*)^B$ when they cross vertically. The edge b has length 1, while b' has length $\sqrt{3}$.

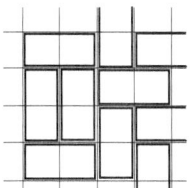

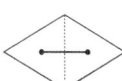

Fig. 2.23 Dimer   **Fig. 2.24** Rectangular tiling

Fig. 2.25 Height function
recovered from a dimer
configuration

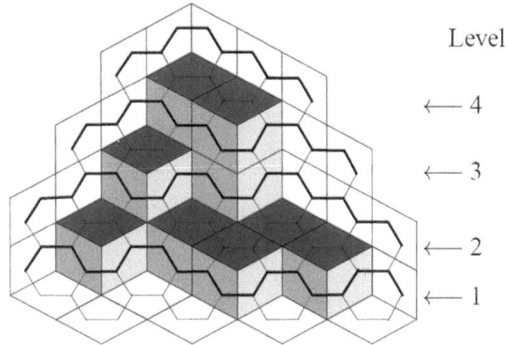

Level

← 4

← 3

← 2

← 1

Dimers on H_N: We call η a *dimer cover* (or a *dimer configuration*) of H_N if η : $H_N^B \to \{0, 1\}$ and the set of sites $\{\{u, v\} \in H_N^B; \eta_b = 1, b = \{u, v\}\}$ covers H_N without overlaps, see the right drawing in Fig. 2.19. The edge satisfying $\eta_b = 1$ is called a *dimer*. Bold segments on hexagons are dimers. In other words, a dimer is a segment connecting the centers of two adjacent equilateral triangles in a triangular lattice, which make a rhombus, see Fig. 2.23.

We can say that a dimer cover η gives a *rhombic tiling* of a triangular lattice without gap. Compare this with a rectangular tiling of a square lattice, see Fig. 2.24.

Now we explain the way to recover the height function from a dimer cover η (in 2D case, this was done by "integrating" the height difference η). Assume that η is given as in the left drawing in Fig. 2.19. We connect non-horizontal dimer edges and horizontal non-dimer edges in this order from the left to the right. This determines one height level. Different height levels do not touch one another and all height levels cover all non-horizontal dimer edges, see Fig. 2.25. The level 1 line has actually height 0.5 (if the side length of the unit box is 1), the level 2 line has height 1.5, and so on.

Let us denote the set of all dimer covers of H_N by $\mathscr{X}_N$. For $\delta = \alpha, \beta, \gamma$ and $\eta \in \mathscr{X}_N$, let us denote the number of δ-edges in η by $n_\delta(\eta)$. Then, for every $\eta \in \mathscr{X}_N$, $n_\alpha(\eta) + n_\beta(\eta) + n_\gamma(\eta) = N^2$ holds. In fact, since a dimer covers H_N, we easily see that $\sharp \text{dimers} \times 2 = \sharp H_N = 2N^2$.

For given N_β, N_γ, we denote by $\mathscr{X}_{N, N_\beta, N_\gamma}$ the family of all $\eta \in \mathscr{X}_N$ such that $n_\delta(\eta) = N_\delta$ for $\delta = \beta, \gamma$. In what follows, we will consider a dynamics η_t which conserves types of edges so that $n_\delta(\eta_t) = N_\delta$ for every $\delta = \alpha, \beta, \gamma$ under the time evolution, where $N_\alpha = N - N_\beta - N_\gamma$. Therefore, $\mathscr{X}_{N,N_\beta,N_\gamma}$ will be the state space of this dynamics. This corresponds to the conservation law of the particle number in particle systems.

Dynamics of dimers on H_N: We introduce a *dimer dynamics* on a honeycomb lattice to describe time evolution of 3D Young diagrams. We express creation and annihilation of a unit box composing a 3D Young diagram by means of dimers. Namely, in Fig. 2.26, types A and B of a dimer configuration on a single hexagon express a *protuberance* and a *cave* in a 3D Young diagram, respectively; see also Fig. 2.19. Then, the generator L of a simple dimer process on H_N is defined as follows: For $f : \mathscr{X}_N \to \mathbb{R}$,

$$Lf(\eta) = \sum_{i \in H_N^*} \left[a 1_{\{\eta_i = A\}} + b 1_{\{\eta_i = B\}} \right] \{ f(\eta^i) - f(\eta) \},$$

where $a, b > 0$, $\eta_i = $ restriction of η to the hexagon i (more precisely, to the six edges of the hexagon with center i), η^i is obtained from η by replacing the type of η_i so that $A \leftrightarrow B$. This generator means that a unit box is created at rate a and annihilated at rate b at every possible position of a 3D Young diagram. Note that, under the transition $\eta \to \eta^i$, $n_\delta(\eta)$ is conserved for every $\delta = \alpha, \beta, \gamma$. Therefore, the configuration space of the dimer process η_t generated by L is $\mathscr{X}_{N,N_\beta,N_\gamma}$.

Let $\mu = \mu_{N,N_\beta,N_\gamma}$ be the uniform probability measure on $\mathscr{X}_{N,N_\beta,N_\gamma}$.

Lemma 2.5. *The invariance of μ under L is equivalent to the condition $a = b$.*

Proof. This is easily seen from the equality

$$\sum_\eta 1_{\{\eta_i = B\}} f(\eta^i) = \sum_\xi 1_{\{\xi_i = A\}} f(\xi),$$

which is obtained by setting $\eta^i = \xi$.   □

Remark 2.2. 1. We can introduce the dimer dynamics also on an infinite region G_∞. Under this dynamics, for the grand canonical ensemble μ_U^ε introduced before Lemma 2.4 to be invariant, the rate of $B \to A$ (creation) is $\varepsilon(< 1)$ while the rate of $A \to B$ (annihilation) is 1 as in the 2D case.

Fig. 2.26 $A \leftrightarrow B$

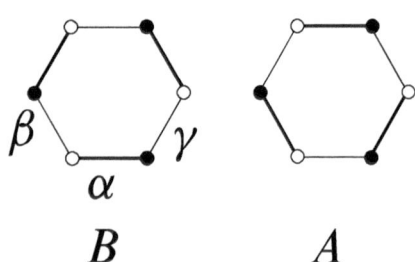

2. Wilson [219] considered a different dynamics and obtained the spectral gap for the generator under a special boundary condition.
3. Caputo, Martinelli, and Toninelli [43] discussed a related dynamics for monotone decreasing height functions defined on $U \Subset \mathbb{Z}^2$, with a fixed boundary condition.

From now on, we assume that $a = b = 1$ and consider the dimer process η_t on $\mathscr{X}_{N,N_\beta,N_\gamma}$ generated by L.

Gibbs measures: Kenyon, Okounkov, and Sheffield [166], and Kenyon [165] proved the following. For $0 \le s, t \le 1$ such that $s + t \le 1$, the limit

$$\mu_{s,t} = \lim_{\frac{N_\beta}{(2\ell+1)^2} \to s, \frac{N_\gamma}{(2\ell+1)^2} \to t} \tilde{\mu}_{G_\ell, N_\beta, N_\gamma},$$

exists, where G_ℓ is a flat honeycomb lattice of size ℓ and $\tilde{\mu}_{G_\ell, N_\beta, N_\gamma}$ is a uniform probability measure on $\mathscr{Y}_{G_\ell, N_\beta, N_\gamma}$, which is defined on G_ℓ similarly to $\mathscr{X}_{G_\ell, N_\beta, N_\gamma}$. The limit $\mu_{s,t}$ is a probability measure on $\mathscr{Y}_\infty = \{$dimer covers of $G_\infty\}$, which is shift invariant (under $e_\alpha, e_\beta, e_\gamma$) and ergodic under shift. $\mu_{s,t}$ satisfies the corresponding DLR equation and has a determinantal structure.

The correlation function of the Gibbs measure decays slowly (only quadratically):

$$\left| \langle \eta_{b_{\delta_1}}; \eta_{b_{\delta_2} + i_\beta e_\beta + i_\gamma e_\gamma} \rangle \right| \le \frac{\text{const}}{|(i_\beta, i_\gamma)|^2},$$

where $\langle \cdot \rangle = \langle \cdot \rangle_{s,t}$ denotes the average under the Gibbs measure and $\langle \eta_{b_1}; \eta_{b_2} \rangle := \langle \eta_{b_1} \eta_{b_2} \rangle - \langle \eta_{b_1} \rangle \langle \eta_{b_2} \rangle$.

Remark 2.3. A similar slow decay of correlation functions is known for the massless Gaussian free field $\phi = \{\phi_i; i \in \mathbb{Z}^d\}$ on $\mathbb{Z}^d$ which has mean 0 and covariance operator $(-\Delta)^{-1}$ (for $d \ge 3$). Indeed, $\langle \phi_i; \phi_j \rangle = G(i, j) \left(= (-\Delta)^{-1}(i, j) \right) \sim \frac{C}{|i-j|^{d-2}}$ as $|i - j| \to \infty$ and, for height differences,

$$\langle (\nabla \phi)_{b_1}; (\nabla \phi)_{b_2} \rangle \sim \frac{c}{\text{dist}(b_1, b_2)^d},$$

is known for $d \ge 2$, see [103] for details, including a more general $\nabla \varphi$-interface model.

The CLT under $\mu_{s,t}$ was established by Kenyon [164] and Boutillier [37]. Indeed, noting that $\sharp G_\ell^B = \left(3(2\ell + 1) \right)^2$, set

$$X_\ell := \frac{1}{3(2\ell + 1)} \sum_{b \in G_\ell^B} (\eta_b - E^{\mu_{s,t}}[\eta_b]).$$

Then, X_ℓ converges in law to a centered Gaussian distribution with variance χ formally given by

$$\chi = \frac{1}{3} \sum_{b_0} \sum_{b \in G_\infty^B} E^{\mu_{s,t}}[\eta_{b_0}; \eta_b],$$

where $\sum_{b_0}$ denotes the sum over three edges which have α, β, γ-types, respectively, and located near the origin 0. If the CLT holds, then the covariance should be given by the limit of

$$E[X_\ell^2] = \frac{1}{\sharp G_\ell^B} \sum_{b_1, b_2 \in G_\infty^B} E^{\mu_{s,t}}[\eta_{b_1}; \eta_{b_2}]$$

$$\sim \frac{1}{\sharp G_\ell^B} \sharp\{b_1 \in G_\ell^B\} \frac{1}{3} \sum_{b_0} \sum_{b_2 \in G_\infty^B} E^{\mu_{s,t}}[\eta_{b_0}; \eta_{b_2}].$$

However, the last sum behaves as

$$\sum_{b_2 \in G_\infty^B} E^{\mu_{s,t}}[\eta_{b_0}; \eta_{b_2}] \sim \sum_{b_2 \in G_\infty^B} \frac{C}{\text{dist}(0, b_2)^2} \sim \int^\infty \frac{C}{r^2} r\,dr = \infty.$$

In particular, χ does not converge absolutely. Nevertheless, the CLT holds. A similar result is known for the $\nabla\phi$-interface model, cf. Naddaf and Spencer [188] and Miller [181] on a bounded region. Recall that the C^2-property of the surface tension is not known for the $\nabla\phi$-interface model and its Hessian is expected to be the CLT variance.

Hydrodynamic limit: Let η_t be the dimer process on $\mathcal{X}_{N,N_\beta,N_\gamma}$ generated by L with $a = b = 1$. For a microscopic dimer configuration $\eta \in \mathcal{X}_N$, we define a macroscopic empirical distribution of δ-edges for $\delta = \beta$ and γ by introducing a spatial scaling as follows:

$$\xi^{\delta,N}(\eta, dx) = \frac{1}{N^2} \sum_{b \in H_N^B : \delta\text{-type}} \eta_b \delta_{\frac{1}{N} x_b}(dx), \quad x = (x_\beta, x_\gamma) \in H,$$

where $x_b = \frac{1}{2}(u+v)$ is the middle point of the edge $b = \{u, v\}$ and H is a continuum torus given by a lozenge with opposite sides identified, see Fig. 2.27.

Introducing a diffusive scaling in time for η_t, the goal is to study the limit of

$$\xi_t^{\delta,N}(dx) = \xi^{\delta,N}(\eta_{N^2 t}, dx),$$

Fig. 2.27 Continuum
lozenge torus

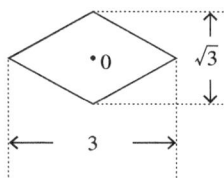

as $N \to \infty$. Our conjecture is that the following result holds: $\xi_t^{\delta,N}(dx)$ converges to $\xi_t^\delta(x)dx$ as $N \to \infty$ and the limit density $\xi_t^\delta(x)$ is characterized as a weak solution of the PDE

$$\frac{\partial \xi_t^\delta}{\partial t} = \frac{\partial}{\partial x_\delta} \left\{ \sum_{\delta_1,\delta_2 \in \{\beta,\gamma\}} D_{\delta_1\delta_2}(\xi_t^\beta, \xi_t^\gamma) \frac{\partial \xi_t^{\delta_2}}{\partial x_{\delta_1}} \right\}, \tag{2.35}$$

where

$$D_{\delta_1\delta_2}(s,t) = \frac{1}{2\chi_{\delta_1\delta_2}} \inf_{g\in\mathscr{C}_0,a_1+a_2=1} \left\langle \left\{ \pi_0 \left(a_1 \sum_i i_\beta \tau_i \eta_{b_\beta} + a_2 \sum_i i_\gamma \tau_i \eta_{b_\gamma} - \Gamma_g \right) \right\}^2 \right\rangle,$$

with

$$\chi_{\delta_1,\delta_2}(s,t) = \sum_i \langle \eta_{b_{\delta_1}} ; \eta_{b_{\delta_2}+i_\beta e_\beta + i_\gamma e_\gamma} \rangle_{s,t},$$

and $\mathscr{C}_0$ is a class of tame functions on $\mathscr{X}$ (the configuration space of dimers on G_∞), τ_i is the shift on G_∞, π_0 is the flip operator $\eta \mapsto \eta^0$, $\Gamma_g = \sum_i \tau_i g$, and $\langle\cdot\rangle_{s,t}$ denotes the average with respect to the Gibbs measures introduced above.

To derive the limit PDE (2.35), we need to compute the time derivative $d\langle \xi_t^{\delta,N}, f \rangle$ for any test function $f \in C^\infty(H)$. We define a *current* W_i at the hexagon $i \in H_N^*$ as

$$W_i(\eta) := 1_{\{\eta_i=A\}} - 1_{\{\eta_i=B\}},$$

that is, W_i is the difference of the rate $1_{\{\eta_i=B\}}$ at which the height function at i increases by 1 and the rate $1_{\{\eta_i=A\}}$ at which the height function at i decreases by 1, or, W_i describes an infinitesimal rate of change of the height function at i. We determine coordinates of H and H_N^* based on e_β and e_α:

$$H = \{x_\beta e_\beta + x_\gamma e_\gamma; x = (x_\beta, x_\gamma) \in \mathbb{T}^2 = (-\tfrac{1}{2}, \tfrac{1}{2}]^2\},$$

$$H_N^* = \{i_\beta e_\beta + i_\gamma e_\gamma; i = (i_\beta, i_\gamma) \in \mathbb{T}_N^2 = N\mathbb{T}^2 \cap \mathbb{Z}^2\}.$$

Fig. 2.28 Definition of $i_1(b)$ and $i_2(b)$

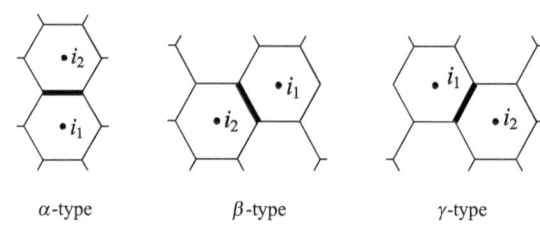

b: α-type β-type γ-type

Then, the drift term of $d\langle \xi_t^{\delta,N}, f\rangle$ is described as follows. $L^N = N^2 L$ is the generator of the time changed process $\eta_{N^2 t}$.

Lemma 2.6. *For every* $\delta = \alpha, \beta, \gamma$, *we have*

$$L^N \langle \xi^{\delta,N}, f\rangle = \frac{1}{N} \sum_{i \in H_N^*} W_i(\eta) \frac{\partial f}{\partial x_\delta}\left(\frac{i}{N}\right) + o\left(\frac{1}{N}\right).$$

Note that the right-hand side is still $O(N)$ since the number of terms in the sum is $O(N^2)$.

Proof. We have

$$L^N \langle \xi^{\delta,N}, f\rangle = \sum_{b:\delta\text{-type}} (L\eta_b) f\left(\frac{x_b}{N}\right).$$

For $b \in H_N^\beta$, let us denote by $i_1(b) \in H_N^*$ the hexagon containing b as its edge and of A-type when $\eta_b = 1$, and by $i_2(b)$ the hexagon containing b as its edge and of B-type when $\eta_b = 1$, see Fig. 2.28. In particular $i_1(b) - i_2(b) = e_\delta$ holds when b is δ-type. Since only $i = i_1(b)$ and $i_2(b)$ remain in the computation of $L\eta_b$, we have

$$L\eta_b = \left[1_{\{\eta_{i_1(b)}=A\}} + 1_{\{\eta_{i_1(b)}=B\}}\right]\left\{(\eta^{i_1(b)})_b - \eta_b\right\}$$
$$+ \left[1_{\{\eta_{i_2(b)}=A\}} + 1_{\{\eta_{i_2(b)}=B\}}\right]\left\{(\eta^{i_2(b)})_b - \eta_b\right\}$$
$$= -W_{i_1(b)} + W_{i_2(b)}.$$

This difference can be translated into the difference of f and we have

$$L^N \langle \xi^{\delta,N}, f\rangle = \sum_{b:\delta\text{-type}} W_{i_1(b)}(\eta) \left\{f\left(\frac{x_b+e_\delta}{N}\right) - f\left(\frac{x_b}{N}\right)\right\}.$$

By taking the first-order term in the Taylor expansion of the difference of f, we obtain the conclusion. □

To kill one more N in the right-hand side of $L^N \langle \xi^{\delta,N}, f \rangle$ given in Lemma 2.6, we need to establish the fluctuation-dissipation relation (sometimes called the gradient replacement due to Varadhan [211]):

$$W_0 \sim \sum_{\delta_1, \delta_2 \in \{\beta, \gamma\}} D_{\delta_1 \delta_2} (\eta_{b_{\delta_1}} - \eta_{b_{\delta_1} + e_{\delta_2}}) + LF,$$

with some $F = F(\eta)$ and a certain diffusion coefficient $D_{\delta_1 \delta_2} = D_{\delta_1 \delta_2}(\xi^\beta, \xi^\gamma)$, where ξ^δ is a limit density of $\xi^{\delta,N}(dx)$ and $\sim$ should be shown with respect to the H^{-1}-norm:

$$\langle\langle h, g \rangle\rangle_{s,t} := \langle (-L)^{-1} h, \Gamma_g \rangle_{s,t}.$$

We consider a Hilbert space $(\mathscr{H}, \langle\langle \cdot, \cdot \rangle\rangle_{s,t})$, where $\mathscr{H} = \overline{L\mathscr{C}_0} \oplus \{W_0\}$ and $\mathscr{C}_0$ is a class of mean 0 functions with respect to $\mu_{s,t}$. To establish the fluctuation-dissipation relation, we need a characterization of closed forms in a certain infinite-dimensional space and the spectral gap of the operator L_{G_ℓ} with an exact order $O(\ell^2)$, note that Wilson's result [219] is only for specific boundary conditions and for a different operator. We note that χ appears as the coefficient of the projection of W_0 to $\nabla_{\delta_2} \eta_{b_\delta}$. For example, taking $\delta_1 = \delta_2 = \beta$, we have

$$\langle\langle \nabla_\beta \eta_{b_\beta}, W_0 \rangle\rangle = \langle (-L)^{-1} \nabla_\beta \eta_{b_\beta}, \Gamma_{W_0} \rangle$$

$$= \langle (-L)^{-1} \nabla_\beta \eta_{b_\beta}, L(\sum_k k_1 \tau_k \eta_{b_\beta}) \rangle$$

$$= -\langle \nabla_\beta \eta_{b_\beta}, \sum_k k_1 \tau_k \eta_{b_\beta} \rangle$$

$$= -\sum_k ((k_1 + 1) - k_1) \langle \eta_{b_\beta}; \eta_{b_\beta + k} \rangle$$

$$= -\chi_{\beta\beta},$$

where the second equality follows by noting that $L\eta_b = -W_{i_1(b)} + W_{i_2(b)}$ and the fourth from the relation $\nabla_\beta \eta_{b_\beta} = \eta_{b_\beta + e_\beta} - \eta_{b_\beta}$.

Remark 2.4. Funaki [101, 102] derived the random motion of a crystallized shape from a microscopic system of interacting Brownian particles in the zero temperature limit. The one-dimensional case [102] is related to the multi-kink motion problem discussed in Sect. 4.2.1.

Chapter 3
Stochastic Partial Differential Equations

So far, we have discussed discrete interface models. Taking their (mesoscopic) continuum limit, as a time evolution of interfaces or some other related physical order parameters, one would expect to obtain stochastic partial differential equations (SPDEs), which are partial differential equations having stochastic terms such as a space-time Gaussian white noise. In connection with the problem of random interfaces, we will especially discuss a stochastic Allen-Cahn equation, sometimes called the time-dependent Ginzburg-Landau (TDGL) equation or a dynamic $P(\phi)$-model, in Chap. 4, and the Kardar-Parisi-Zhang (KPZ) equation in Chap. 5. This chapter explains, taking the TDGL equation as an example, some fundamental facts concerning SPDEs such as white noises, colored noises, definitions and regularities of solutions of the SPDEs we are interested in. We give some other examples of related SPDEs. The SPDEs used in physics are sometimes ill-posed. They appear in a wide variety of fields, not only in physics but also in biology, engineering (e.g., in control theory, filtering), economics (e.g., in mathematical finance), and others.

3.1 The TDGL Equation

Here, we start with the equation for a stochastic quantization or a dynamic $P(\phi)_d$-model taking $P(\phi) = \frac{1}{4}\phi^4$:

$$\partial_t \phi = \Delta\phi - \phi^3 + \dot{W}(t, x), \quad x \in \mathbb{R}^d, \tag{3.1}$$

where $\dot{W}(t, x)$ is a *space-time (Gaussian) white noise*, that is, a Gaussian system with mean 0 and the covariance structure

$$E[\dot{W}(t, x)\dot{W}(s, y)] = \delta(t - s)\delta(x - y), \quad t, s \geq 0, \ x, y \in \mathbb{R}^d. \tag{3.2}$$

Expression (3.2) is formal and a more precise explanation will be given in Sect. 3.2 later. It is known that the noise is very irregular: $\dot{W} \in C^{-\frac{d+1}{2}-} := \bigcap_{\delta>0} C^{-\frac{d+1}{2}-\delta}$

© The Author(s) 2016

T. Funaki, *Lectures on Random Interfaces*, SpringerBriefs in Probability and Mathematical Statistics, DOI 10.1007/978-981-10-0849-8_3

a.s. The solution, at least in the linear case (without ϕ^3), has a better regularity: $\phi(t,x) \in C^{\frac{2-d}{4}-,\frac{2-d}{2}-}$ (a.s.) because of the regularization property of the Laplacian Δ, see Proposition 3.1 below. This implies ϕ can be an ordinary function only when $d = 1$, so that the nonlinear equation (3.1) is well-posed only when $d = 1$.

In the higher-dimensional case, the SPDE (3.1) is *ill-posed*. We will give a very quick overview of the approaches to such ill-posed SPDEs using regularity structures initiated by Hairer and the paracontrolled calculus introduced by Gubinelli and others later in this section and also in Sect. 5.1 especially for the KPZ equation.

The equation (3.1) is sometimes called the *time-dependent Ginzburg-Landau* *(TDGL) equation*. The TDGL equation is a very important model equation in physics. It has, for $\mathbb{R}$-valued (sometimes $\mathbb{C}$-valued) $u = u(t,x)$, the form

$$\partial_t u = -\frac{1}{2}(-\Delta)^\alpha \frac{\delta H}{\delta u(x)}(u) + (-\Delta)^{\alpha/2}\dot{W}(t,x), \quad x \in \mathbb{R}^d, \tag{3.3}$$

where $\dot{W}(t,x)$ is the space-time Gaussian white noise and $H(u)$ is a (formal) Hamiltonian, called the Ginzburg-Landau-Wilson free energy, given by

$$H(u) = \int_{\mathbb{R}^d} \left\{ \frac{1}{2}|\nabla u(x)|^2 + V(u(x)) \right\} dx, \tag{3.4}$$

with a self-potential $V : \mathbb{R} \to \mathbb{R}$, see Hohenberg and Halperin [146], and also van Saarloos and Hohenberg [209], van Saarloos [210]. This is a kind of Langevin equation, and the Gibbs measure formally given by $\frac{1}{Z}e^{-H(u)}du$, $du = \prod_{x \in \mathbb{R}^d} du(x)$, is invariant under the corresponding dynamics, see Sect. 3.3.3 for more details. When $\alpha = 0$, the equation (3.3) is called Model A (non-conservative system), whereas when $\alpha = 1$, it is called Model B (conservative system). In the latter case, the integral $\int_{\mathbb{R}^d} u(t,x)dx$ is conserved under the time evolution, at least at heuristic level, as we explain below. This integral represents the total mass, the total volume or other quantities depending on the situation we consider.

The functional derivative of H is given by

$$\frac{\delta H}{\delta u(x)} = -\Delta u(x) + V'(u(x)).$$

In fact, for every test function $\varphi \in C_0^\infty(\mathbb{R})$, we have

$$\frac{d}{d\varepsilon}H(u + \varepsilon\varphi)\Big|_{\varepsilon=0} = \int_{\mathbb{R}^d} \frac{d}{d\varepsilon}\left\{ \frac{1}{2}|\nabla u(x) + \varepsilon\nabla\varphi(x)|^2 + V(u(x) + \varepsilon\varphi(x)) \right\}\Big|_{\varepsilon=0} dx$$

$$= \int_{\mathbb{R}^d} \left\{ \nabla u(x) \cdot \nabla\varphi(x) + V'(u(x))\varphi(x) \right\} dx$$

$$= \int_{\mathbb{R}^d} \left\{ -\Delta u(x) + V'(u(x)) \right\} \varphi(x)dx.$$

Therefore, the SPDE (3.3) has the forms

$$\partial_t u = \frac{1}{2}\Delta u - \frac{1}{2}V'(u) + \dot{W}(t, x),$$ (3.5)

in the case $\alpha = 0$ and

$$\partial_t u = -\frac{1}{2}\Delta^2 u + \frac{1}{2}\Delta\{V'(u)\} + \sqrt{-\Delta}\,\dot{W}(t, x),$$ (3.6)

in the case $\alpha = 1$. Note that $\sqrt{-\Delta}\,\dot{W}(t, x)$ and $\mathrm{div}\,\dot{\mathbb{W}}(t, x)$ have the same covariance structure, where $\dot{\mathbb{W}}(t, x) = \{\dot{W}^i(t, x)\}_{i=1}^d$ is a family of d independent copies of the space-time white noise. The SPDEs (3.5) and (3.6) are called the *TDGL equation of non-conservative type* and *of conservative type*, respectively. Indeed, (3.6) has a "formal" conservation law: Consider on $\mathbb{T}^d$ instead of $\mathbb{R}^d$. Then, at least heuristically,

$$\int_{\mathbb{T}^d}\Delta\{-\Delta u(x) + V'(u(x))\}dx = \int_{\mathbb{T}^d}\Delta 1 \cdot \{-\Delta u(x) + V'(u(x))\}dx = 0,$$

$$\int_{\mathbb{T}^d}\mathrm{div}\,\dot{\mathbb{W}}dx = -\int_{\mathbb{T}^d}\nabla 1 \cdot \dot{\mathbb{W}}dx = 0,$$

and, from these, we have $\partial_t \int_{\mathbb{T}^d} u(t, x)dx = 0$, which implies that $\int_{\mathbb{T}^d} u(t, x)dx$ is conserved in time.

If V is a double-well potential $V(u) = \frac{1}{4}(u^2 - 1)^2$ with two bottoms at $u = \pm 1$ of same depth, then $-V'(u) = u - u^3$ gives a bistable reaction term, i.e., $u = \pm 1$ are stable, while $u = 0$ is unstable, see Fig. 3.1. When $\dot{W} = 0$ (i.e., without the noise term), (3.5) is known as the Allen-Cahn equation [6] or the reaction-diffusion equation of bistable type, while (3.6) is known as the Cahn-Hilliard equation.

The *sharp interface limit*, or the problem of the *dynamic phase transition*, is to study the limit as $\varepsilon \downarrow 0$ of the TDGL equation of non-conservative type (or a stochastic Allen-Cahn equation)

$$\partial_t u = \Delta u + \frac{1}{\varepsilon^2}f(u) + \dot{W}^\varepsilon(t, x), \quad x \in \mathbb{R}^d,$$ (3.7)

Fig. 3.1 Double-well potential

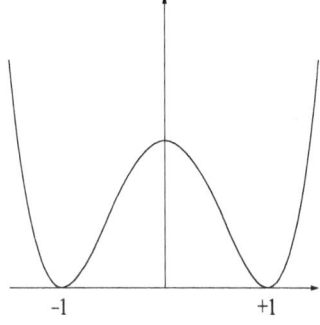

Fig. 3.2 Phase separation

where $f = -V'$ with a potential V of double-well type, e.g., $f(u) = u - u^3$, and the noise term $\dot{W}^\varepsilon$ requires some scaling. The limit is expected to satisfy

$$u(t,x) \xrightarrow[\varepsilon \downarrow 0]{} \begin{cases} +1 \\ -1 \end{cases}.$$

In other words, a random phase separating hyperplane Γ_t appears and the problem is to determine its dynamics under a proper time scaling, see Fig. 3.2. A similar problem is discussed also for the stochastic Cahn-Hilliard equation. These are the main topics discussed in Chap. 4.

The SPDEs (3.3) with reflection (i.e., u is confined to be $u \geq 0$ or stay between two walls) are studied by Nualart and Pardoux [191], Funaki and Olla [110] for Model A and by Debussche and Zambotti [59] for Model B.

If the self-potential V has a discontinuity such as $V(u) = \beta 1_{[0,\infty)}(u)$, then $V'(u) = \beta \delta_0(u)$ and the SPDE (3.5) admits a singular drift:

$$\partial_t u = \frac{1}{2}\Delta u - \frac{\beta}{2}\delta_0(u) + \dot{W}(t,x). \tag{3.8}$$

This potential physically means that the space is filled by two different media separated by a membrane located at the level $u = 0$. Note that J in (1.1) has essentially the same form as H in (3.4) in this case. If $\beta > 0$, the solution u of (3.8) is pushed toward the negative side with strength β when it touches the membrane located at 0. Bounebache and Zambotti [36] studied such an SPDE with singular drift. This is a continuous analog of the Ginzburg-Landau (i.e., an evolutional version of) $\nabla\varphi$-interface model in two media.

As we pointed out, the SPDE (3.3) is well-posed only when the spatial dimension is $d = 1$. For higher dimensions, instead, one can consider the corresponding spatially discretized equation. The Ginzburg-Landau $\nabla\varphi$-interface model is described by such an equation, see Funaki and Spohn [116], Funaki [103]. We now discuss approaches to ill-posed SPDEs.

Regularity structures and paracontrolled calculus: Hairer [139] introduced the theory of regularity structures and a systematic way for the renormalization for ill-posed SPDEs, see Friz and Hairer [88] and Chandra and Weber [47] for overviews. Another approach was given by Gubinelli, Imkeller, and Perkowski [130] based on the paracontrolled calculus. The SPDEs for which these theories are applicable include the dynamic ϕ_d^4-model (3.1) with $d = 2, 3$ (on a torus $\mathbb{T}^d$ or on $\mathbb{R}^d$), the KPZ equation (5.1) in one-dimension, the continuous parabolic Anderson model described by $\partial_t u = \Delta u + u\dot{W}(x)$ with spatial white noise $\dot{W}(x)$ when $d = 2, 3$

(if $d = 1$, this equation is well-posed), and the stochastic Navier-Stokes equation with space-time white noise (local existence and uniqueness on $\mathbb{T}^3$ was shown by Zhu and Zhu [224]).

For (3.1), $d = 2, 3$ is subcritical (under certain scaling, see Definition 1.3 of [47]) and the equation is renormalizable, while $d = 4$ is critical and $d \geq 5$ are supercritical, so that (3.1) is not renormalizable if $d \geq 4$. The renormalization procedure goes as follows. We replace $\dot{W}$ by a smeared noise $\dot{W}^\varepsilon$ and at the same time replace ϕ^3 by $\phi^3 - C_\varepsilon \phi$, with a renormalization constant C_ε depending on ε and d. Then, choosing $C_\varepsilon \uparrow \infty$ properly, a non-trivial limit of $\phi = \phi^\varepsilon$ as $\varepsilon \downarrow 0$ exists (locally in time). Actually, the global well-posedness was shown by Mourrat and Weber [184] for the dynamic ϕ_2^4-model on $\mathbb{R}^2$ and by Mourrat and Weber [185] for the dynamic ϕ_3^4-model on $\mathbb{T}^3$. The solution obtained in the limit is not continuous in the noise ξ (in place of $\dot{W}^\varepsilon$), but roughly speaking continuous in ξ and (finitely many) polynomials in it constructed probabilistically (like Wick products) under a suitable topology. The solution map is therefore factorized into two parts.

The convergence to the dynamic ϕ_2^4-model on $\mathbb{T}^2$ starting from the microscopic particle model called the Ising-Kac model was shown by Mourrat and Weber [183]. The convergence of the discretized dynamic ϕ_3^4-model on $\mathbb{T}^3$ was studied by Hairer and Matetski [140], who also developed a general framework for discretization in the setting of regularity structures.

SPDEs with multiplicative noise of the form $\partial_t u = \partial_x^2 u + H(u) + G(u)\dot{W}(t, x)$ were discussed by Hairer and Pardoux [142].

For the KPZ equation, $d = 2$ is critical and $d = 1$ is subcritical, so that the KPZ equation is renormalizable only when $d = 1$, see Chap. 5.

Other SPDEs: The fluctuation limit in the hydrodynamic limit for microscopic interacting (particle) systems leads to SPDEs. Related fluctuation-dissipation theorem or Green-Kubo formula are studied. Recall Sect. 2.2.3 and see Funaki and Olla [110]. See also Funaki [99] for the related hydrodynamic limit.

The random motions of curves (or strings) in $\mathbb{R}^d$ (Funaki [89]), curves (or loops) on a manifold M (Funaki [94]), and curves in a convex domain D with reflection (Bounebache [35]) are studied by means of $\mathbb{R}^d$ or M or D-valued SPDEs, respectively.

3.2 White Noise, Colored Noise, and Stochastic Integrals

To construct the space-time Gaussian white noise $\dot{W}(t, x)$ having the covariance structure (3.2), take a complete orthonormal system (CONS) $\{\psi_k\}_{k=1}^\infty$ in $L^2(\mathbb{R}^d, dx)$ and a system of independent one-dimensional Brownian motions $\{B_t^k\}_{k=1}^\infty$ realized on a certain probability space $(\Omega, \mathscr{F}, P)$, and consider a formal Fourier series with random coefficients:

$$W(t, x) = \sum_{k=1}^\infty B_t^k \psi_k(x), \qquad (3.9)$$

which is called the *white noise process* or *cylindrical Brownian motion* on $L^2(\mathbb{R}^d)$. Then, one would have that

$$E[W(t, x)W(s, y)] = \sum_{k,j=1}^{\infty} E[B_t^k B_s^j]\psi_k(x)\psi_j(y)$$

$$= \sum_{k=1}^{\infty} (t \wedge s)\psi_k(x)\psi_k(y) = (t \wedge s)\delta(x - y),$$

and thus its time derivative is expected to satisfy the relation (3.2). In fact, $\frac{\partial}{\partial t}(t \wedge s) = 1_{[0,s)}(t) = 1_{(t,\infty)}(s)$, so that $\frac{\partial}{\partial s}\frac{\partial}{\partial t}(t \wedge s) = \delta(t - s)$. As an example of the CONS $\{\psi_k\}_{k=1}^{\infty}$, one can take the Hermite functions

$$h_k(x) = \left(2^k k! \sqrt{\pi}\right)^{-\frac{1}{2}} e^{-\frac{x^2}{2}} H_k(x),$$

where $H_k(x)$ are the Hermite polynomials: $H_k(x) = (-1)^k e^{x^2} \frac{d^k}{dx^k} e^{-x^2}$. One can also formally construct $\dot{W}(t, x)$ as $\dot{W}(t, x) = \dot{B}(t)\dot{W}^1(x_1) \cdots \dot{W}^d(x_d)$ with independent (two-sided) Brownian motions $B, W^1, \ldots, W^d$.

Stochastic integrals with respect to $W(t, x)$ are well-defined as follows. For $(\mathscr{F}_t)$-adapted $f = f(t, x, \omega) \in L^2([0, T] \times \mathbb{R}^d \times \Omega)$ (for every $T > 0$), set

$$M_t(f) \equiv \int_0^t \int_{\mathbb{R}^d} f(s, x)W(dsdx) := \sum_{k=1}^{\infty} \int_0^t \left(f(s, \cdot), \psi_k\right)_{L^2(\mathbb{R}^d)} dB_s^k, \qquad (3.10)$$

where $\mathscr{F}_t = \sigma\{W(s, \cdot); s \le t\}$ and each term in the right-hand side is defined as a usual stochastic integral with respect to B_t^k. Then, $(M_t(f))$ is an $(\mathscr{F}_t)$-martingale and the *Itô isometry* holds:

$$\|M_T(f)\|_{L^2(\Omega)} = \|f\|_{L^2([0,T]\times\mathbb{R}^d\times\Omega)},$$

that is,

$$E[M_T^2(f)] = \int_0^T \int_{\mathbb{R}^d} E[f(t, x)^2]dtdx.$$

In fact, one can easily calculate

$$E[M_T^2(f)] = E\left[\left\{\sum_{k=1}^{\infty} \int_0^T \left(f(t, \cdot), \psi_k\right)_{L^2(\mathbb{R}^d)} dB_T^k\right\}^2\right]$$

$$= \sum_{k=1}^{\infty} E\left[\left\{\int_0^T (f(t,\cdot),\psi_k)_{L^2(\mathbb{R}^d)}\, dB_t^k\right\}^2\right]$$

$$= \sum_{k=1}^{\infty} \int_0^T E\left[(f(t,\cdot),\psi_k)_{L^2(\mathbb{R}^d)}^2\right] dt$$

$$= \int_0^T E\left[\|f(t,\cdot)\|_{L^2(\mathbb{R}^d)}^2\right] dt,$$

where the second line follows since the stochastic integrals with independent Brownian motions are orthogonal in $L^2(\Omega)$, third line is due to the Itô isometry for the stochastic integrals with respect to one-dimensional Brownian motions, and the last line follows from the Plancherel equality. In particular, for $\psi \in L^2(\mathbb{R}^d)$,

$$W(t,\psi) := M_t(\psi) \equiv \int_0^t \int_{\mathbb{R}^d} \psi(x) W(dsdx) \left(= \int_0^t \int_{\mathbb{R}^d} \psi(x)\dot{W}(s,x)dsdx\right)$$

has a meaning and, by the Itô isometry, we have

$$E[W(t,\psi)^2] = t\|\psi\|_{L^2(\mathbb{R}^d)}^2,$$

and thus $W(t,\psi)/\|\psi\|_{L^2(\mathbb{R}^d)}$ is a one-dimensional Brownian motion for each ψ. The *Burkholder-Davis-Gundy's inequality* can be easily extended to this setting: For every $p > 0$, there exists $C = C_p > 0$ such that

$$E\left[\sup_{0 \le t \le T} |M_t(f)|^p\right] \le C E\left[\left(\int_0^T \int_{\mathbb{R}^d} f(t,x)^2 dtdx\right)^{p/2}\right].$$

Unfortunately, although $M_t(f)$ in (3.10) is well-defined, the series (3.9) does not converge in the space $L^2(\mathbb{R}^d)$, so that the cylindrical Brownian motion does not stay on this space. Instead, if we add a damping factor $\{\lambda_k > 0\}_{k=1}^{\infty}$ satisfying $\operatorname{Tr} Q \equiv \sum_{k=1}^{\infty} \lambda_k < \infty$ and consider

$$W^Q(t,x) = \sum_{k=1}^{\infty} \sqrt{\lambda_k} B_t^k \psi_k(x), \qquad (3.11)$$

then it converges in $L^2(\mathbb{R}^d)$; indeed, $E[\|W^Q(t,\cdot)\|_{L^2(\mathbb{R}^d)}^2] = E[\sum_{k=1}^{\infty}(\sqrt{\lambda_k}B_t^k)^2] = t\operatorname{Tr} Q < \infty$, and the limit $W^Q(t,x)$ is called a *Q-Brownian motion*, where Q is a linear operator on $L^2(\mathbb{R}^d)$ such that $Q\psi_k = \lambda_k \psi_k$. Its time derivative is called a *colored noise* and has the covariance structure

$$E[\dot{W}^Q(t,x)\dot{W}^Q(s,y)] = \delta(t-s)Q(x,y), \qquad (3.12)$$

where $Q(x, y) = \sum_{k=1}^{\infty} \lambda_k \psi_k(x) \psi_k(y)$. Note that Q has a representation as an integral operator of trace class $Q\psi(x) = \int_{\mathbb{R}^d} Q(x, y)\psi(y)dy$ with the kernel $Q(x, y)$ satisfying $\operatorname{Tr} Q = \int_{\mathbb{R}^d} Q(x, x)dx < \infty$. Note also that, for $Q = I$ (i.e., for the cylindrical Brownian motion), $\operatorname{Tr} I = \infty$.

If we consider $q(x, y) = \sum_{k=1}^{\infty} \sqrt{\lambda_k} \psi_k(x) \psi_k(y)$, then

$$W^Q(t, x) = \int_0^t \int_{\mathbb{R}^d} q(x, y)W(dsdy),$$

or its time derivative has the covariance structure (3.12), where W is the white noise process. The operator $\sqrt{Q}$ defined by $\sqrt{Q}\psi(x) = \int_{\mathbb{R}^d} q(x, y)\psi(y)dy$ is of Hilbert-Schmidt type with the kernel $q(x, y) \in L^2(\mathbb{R}^d \times \mathbb{R}^d)$ and $(\sqrt{Q})^2\psi(x) = Q\psi(x)$, since $\int_{\mathbb{R}^d} q(x, y)q(y, z)dy = Q(x, z)$.

See the book of Da Prato and Zabczyk [57] or [104] written in Japanese for more systematic discussions on Brownian motions on general separable Hilbert spaces.

3.3 SPDEs of Parabolic Type with Additive Noises

We consider the TDGL equations (3.5) of non-conservative type and (3.6) of conservative type in extended forms. This section (except for Sect. 3.3.3) is mostly taken from Funaki [93].

Let us consider the following SPDEs of parabolic type with additive noises for $u = u(t, x)$, $t \geq 0$, $x \in \mathbb{R}^d$:

$$\partial_t u = Au + B(u) + C\dot{W}(t, x), \tag{3.13}$$

with the space-time Gaussian white noise $\dot{W}(t, x)$. Here, $A = \sum_{|\alpha| \leq 2m} a_\alpha(x)D^\alpha$ is a $2m$th-order differential operator satisfying the so-called uniform ellipticity condition (see [93]), $B(u)$ is a nonlinear term of the form $B(u)(x) = B\{b(x, u)\}$ with $B = \sum_{|\alpha| \leq n} b_\alpha(x)D^\alpha$ and a nonlinear functional $b(x, u)$ of u, and $C = \sum_{|\alpha| \leq \ell} c_\alpha(x)D^\alpha$ is an ℓth-order differential operator, where $D^\alpha = \left(\frac{\partial}{\partial x_1}\right)^{\alpha_1} \cdots \left(\frac{\partial}{\partial x_d}\right)^{\alpha_d}$, $|\alpha| = \sum_{i=1}^d \alpha_i$ for $\alpha = (\alpha_1, \ldots, \alpha_d) \in \mathbb{Z}_+^d$ and the coefficients satisfy $a_\alpha, b_\alpha, c_\alpha \in C_b^\infty(\mathbb{R}^d)$.

If we consider the Hamiltonian

$$H(u) = \int_{\mathbb{R}^d} \left\{ \frac{1}{2} \mathscr{A} u(x) \cdot u(x) + V(u(x)) \right\} dx,$$

instead of (3.4) and if $\mathscr{A}$ has a symmetric form:

$$\mathscr{A} u(x) = \sum_{|\alpha|, |\beta| \leq m} (-1)^{|\alpha|} D^\alpha \{ a_{\alpha\beta} D^\beta u \}(x)$$

with coefficients satisfying $a_{\alpha\beta} = a_{\beta\alpha}$ and $\{a_{\alpha\beta}\}_{|\alpha|=|\beta|=m}$ being positive definite, then we have

$$\frac{\delta H}{\delta u(x)} = \mathscr{A}u(x) + V'(u(x)).$$

The corresponding TDGL equation has a solution with better regularity for larger m. Note that $m = 1, n = 0, \ell = 0$ for the TDGL equation (3.5) of non-conservative type, while $m = 2, n = 2, \ell = 1$ for the TDGL equation (3.6) of conservative type.

3.3.1   Two Definitions of Solutions

We take the weighted L^2-spaces

$$L_r^2 = L^2(\mathbb{R}^d, e^{-2r\chi(x)}dx), \quad r > 0,$$

over $\mathbb{R}^d$ as the state spaces for solutions of (3.13), where $\chi \in C^\infty(\mathbb{R}^d)$ is such that $\chi(x) = |x|$ for $|x| \geq 1$.

Definition 3.1. $u(t,x)$ is called a *solution* of (3.13) with an initial value $u_0 \in L_r^2, r > 0$ *in the sense of generalized functions*, if it satisfies

$$\langle u(t), \varphi \rangle = \langle u_0, \varphi \rangle + \int_0^t \{\langle u(s), A^*\varphi \rangle + \langle b(\cdot, u(s)), B^*\varphi \rangle\}ds + \langle W(t), C^*\varphi \rangle,$$
$$(3.14)$$

for all $\varphi \in C_0^\infty(\mathbb{R}^d)$, where $\langle u, \varphi \rangle = \int_{\mathbb{R}^d} u(x)\varphi(x)dx$ and A^*, B^*, C^* denote the adjoints of A, B, C, respectively.

We have multiplied (3.13) by the test function φ and integrated it with respect to t and x. Then, we formally obtain (3.14) by regarding

$$\int_0^t ds \int_{\mathbb{R}^d} C\dot{W}(s,x)\varphi(x)dx = \int_0^t \int_{\mathbb{R}^d} C^*\varphi(x)W(dsdx) = \langle W(t), C^*\varphi \rangle,$$

which is defined as a stochastic integral. Another way to give a mathematical meaning to (3.13) is via the Duhamel's principle.

Definition 3.2. $u(t,x)$ is called a *mild solution* of (3.13), if it satisfies

$$u(t,x) = \int_{\mathbb{R}^d} p(t,x,y)u_0(y)dy + \int_0^t \int_{\mathbb{R}^d} B_y^* p(t-s,x,y)b(y,u(s,y))dsdy$$

$$+ \int_0^t \int_{\mathbb{R}^d} C_y^* p(t-s,x,y)W(dsdy), \qquad (3.15)$$

where $p(t, x, y)$ is the fundamental solution of the parabolic operator $\partial_t - A$ defined in the space L_r^2. The last term in (3.15) is defined as a stochastic integral.

In typical cases, working in an appropriate space, these two notions of solutions are equivalent, see [104, 117]. If $b(x, u)$ is Lipschitz continuous as a map from $u \in L_r^2$ to $b(\cdot, u) \in L_r^2$, the (mild) solution of the SPDE (3.13) exists uniquely under the conditions on m, n, ℓ in Proposition 3.1 stated below. One can apply the standard method of successive approximations. Also, pathwise uniqueness holds.

3.3.2 Regularity of Solutions

As we pointed out, the noise $W(t, x)$ lives in a bad space and we need the regularizing properties of the operator A. Recall that $C^{\alpha-, \beta-} := \bigcap_{\delta > 0} C^{\alpha - \delta, \beta - \delta}$.

Proposition 3.1. *Assume* $2m > 2\ell + d$ *and* $2m > n$. *Then, for the solution* $u(t, x)$ *of* (3.13), *we have that*

$$u(t, x) \in C^{\alpha-, \beta-}((0, \infty) \times \mathbb{R}^d), \quad \text{a.s.,} \tag{3.16}$$

with

$$\alpha = \frac{2m - 2\ell - d}{4m} \quad \text{and} \quad \beta = \frac{2m - 2\ell - d}{2}.$$

In particular, for the TDGL equation (3.5) of non-conservative type, we have that $u(t, x) \in C^{\frac{2-d}{4}-, \frac{2-d}{2}-}((0, \infty) \times \mathbb{R}^d)$, and for the TDGL equation (3.6) of conservative type, we have that $u(t, x) \in C^{\frac{2-d}{8}-, \frac{2-d}{2}-}((0, \infty) \times \mathbb{R}^d)$. Therefore, the solutions live in the usual function spaces only when $d = 1$. If $d = 2$, the solutions are already generalized functions. Of course, this can be improved, if we take ℓ to be negative, which results in a more regular noise in the space variable, that is, a colored noise. Otherwise, the solution cannot be defined in a classical sense.

Outline of the proof of Proposition 3.1. The details are given in [93], which can be actually improved as we formulated in the statement of the proposition, without much difficulty, as we explain below. Here we discuss the case with $u_0 = 0$ and $B = 0$ only, for which the SPDE (3.13) is linear and the solution is an Ornstein-Uhlenbeck process in an infinite-dimensional space. The following estimate is known for the fundamental solution $p(t, x, y)$ introduced in Definition 3.2:

$$\left| \partial_t^j D_x^\alpha D_y^\beta p(t, x, y) \right| \leq t^{-\frac{|\alpha| + |\beta|}{2m} - j} \bar{p}(t, x, y), \quad t \in (0, T], \; x, y \in \mathbb{R}^d,$$

where

$$\bar{p}(t, x, y) = K_1 t^{-\frac{d}{2m}} \exp\left\{-K_2 \left(\frac{|x-y|^{2m}}{t}\right)^{\frac{1}{2m-1}}\right\}, \quad t \in (0, T], \ x, y \in \mathbb{R}^d.$$

As we mentioned above, we consider the regularity of the last term in the right-hand side of (3.15), which we call $u^W(t, x)$. Then, using the Itô isometry and the estimate on the fundamental solution, one can show that

$$E\left[\left|D^\alpha u^W(t, x) - D^\alpha u^W(t', x')\right|^2\right]$$

$$\leq C\left\{|t - t'|^{\frac{2m-2l-d-2|\alpha|}{2m}} + |x - x'|^{(2m-2l-d-2|\alpha|-\delta)\wedge 2}\right\},$$

$$t, t' \in (0, T], \ x, x' \in \mathbb{R}^d, \ \delta > 0,$$

as long as both exponents are positive. Noting that $u^W(t, x)$ is Gaussian (so that estimates on higher moments can be deduced from the above L^2-estimate) and applying the Kolmogorov-Chentsov theorem (see, for example, Kunita [172]), we obtain (3.16) for $u^W(t, x)$. Other terms in (3.15) have better regularity at least if $n < 2m$, see [93] for details.   □

3.3.3   Invariant Measures

Once the existence, uniqueness, and regularity of solutions are established, the next interesting question is to study their asymptotic behavior as the time t becomes large, in particular, the existence and uniqueness of invariant measures and the ergodicity. For finite-dimensional Markov processes, one can apply the well-known lower bound technique due to Doeblin. However, the problem becomes delicate for processes taking values in infinite-dimensional spaces. The solutions of SPDEs take values in certain Polish spaces. In order to investigate the above mentioned properties for such processes, the following general methods are known: (1) *strong Feller property* (the corresponding semigroup P_t has the property $P_t\varphi \in C_b$ for all $\varphi \in \mathcal{B}_b$, Da Prato and Zabczyk [58]), (2) *asymptotic strong Feller property* (the transition measures $P(t, x, \cdot)$ and $P(t, y, \cdot)$ become close as $t \to \infty$ for every x and y, Hairer and Mattingly [141], Hairer [137]), and (3) *e-property* (Komorowski, Peszat, and Szarek [171]). These methods are applied to various types of SPDEs.

Funaki [92] studied the invariant or reversible measures of the TDGL equation. It is shown that the (grand canonical) Gibbs measure associated with the Hamiltonian $H(u)$ is reversible under the TDGL equation (3.5) of non-conservative type and the uniqueness of a (tempered) invariant measure is shown under the assumption that the potential V is convex. In contrast, the reversible measures of the TDGL equation (3.6) of conservative type are not unique. This equation has a family of canonical Gibbs measures (Gibbs measures associated with the Hamiltonian $H_{\lambda(\cdot)} = H(u) - \int \lambda(x)u(x)dx$ with external fields λ which satisfy $\Delta\lambda = 0$) as its reversible measures. This fact is essential for discussing the hydrodynamic limit, see [90, 91].

Chapter 4
Sharp Interface Limits for a Stochastic Allen-Cahn Equation

This chapter discusses the problem of the sharp interface limits for a stochastic Allen-Cahn equation as mentioned in Sect. 3.1. The original motivation comes from statistical physics in studying the dynamic phase transition [161, 162]. The proper scaling in time is different and changes according to the types of noises and the spatial dimension d. We start with explaining the results in a deterministic case and then formulate the results in a stochastic case. A brief survey is given on the motions by mean curvature with or without noises, which arise in the limit, and the sharp interface limit.

4.1 Setting, Quick Overview and Background

Let us consider the *stochastic Allen-Cahn equation* (stochastic reaction-diffusion equation or modified TDGL equation) with a small parameter $\varepsilon > 0$:

$$\partial_t u = \Delta u + \frac{1}{\varepsilon^2} f(u) + \dot{W}^\varepsilon(t, x), \quad t > 0, \ x \in D, \tag{4.1}$$

where $\dot{W}^\varepsilon(t, x)$ is a space-time noise depending on ε, which will be specified later, and D is a domain in $\mathbb{R}^d$. We require some boundary conditions. In the TDGL equation (3.5) of non-conservative type, the noise was the space-time Gaussian white noise and the nonlinear reaction term was given by $-\frac{1}{2} V'(u)$, but in (4.1) we take $\frac{1}{\varepsilon^2} f(u)$ with $f = -V'$ instead. We also drop $\frac{1}{2}$ in front of Δ for simplicity. In this chapter, we will treat some other noises as well.

We assume the reaction term $f \in C^\infty(\mathbb{R})$ is bistable, that is,

$$f(\pm 1) = f(u_*) = 0, \ f'(\pm 1) < 0, \ f'(u_*) > 0, \tag{4.2}$$

with some $u_* \in (-1, 1)$, see Fig. 4.1. A typical example is $f(u) = u - u^3$. As we mentioned in Sect. 3.1, we would expect that $\lim_{\varepsilon \downarrow 0} u^\varepsilon(t, x) = +1$ or -1 for the

© The Author(s) 2016
T. Funaki, *Lectures on Random Interfaces*, SpringerBriefs in Probability and Mathematical Statistics, DOI 10.1007/978-981-10-0849-8_4

Fig. 4.1 Bistable reaction
term

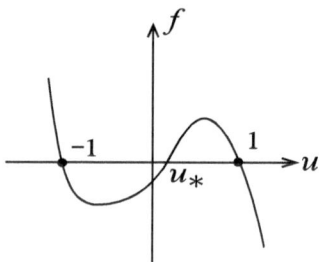

solution $u = u^\varepsilon(t, x)$ of (4.1), since ± 1 are stable points of f (or of the reaction
dynamical system $\dot{u} = f(u)$). Note that u_* is an unstable point of f. The goal is to
find the evolution law of the random interface Γ_t separating the two stable phases
± 1, see Fig. 3.2.

Before formulating the results on the sharp interface limit for the stochastic
Allen-Cahn equation (4.1), we give a quick overview:

1. $d = 1$ (Funaki [96, 97]): Replace $\dot{W}(t, x)$ by a small space-time Gaussian white
   noise $\varepsilon^\gamma \dot{W}(t, x)$ with $\gamma > \frac{19}{2}$. Recall that with such noise the SPDE (4.1) is well-
   posed only in one-dimension. Then, under a proper initial condition (in particular,
   with a single interface), we can show that

   $$u^\varepsilon(t, x) := u(\varepsilon^{-2\gamma-1}t, x) \longrightarrow \chi_{\xi_t}(x) = 1_{(-\infty,\xi_t)} - 1_{(\xi_t,\infty)},$$

   as $\varepsilon \downarrow 0$, where the phase-separating point ξ_t behaves as a Brownian motion
   multiplied by a constant called the inverse surface tension. The meaning of the
   condition $\gamma > \frac{19}{2}$ will be explained later.
2. $d \geq 2$ (Funaki [98], Weber [216]): Take $\dot{W}^\varepsilon(t, x) \equiv \dot{W}^\varepsilon(t) = \frac{1}{\varepsilon}\dot{\Xi}^\varepsilon(t)$ with
   $\Xi^\varepsilon(t) \sim \dot{W}_t$ (time-dependent white noise) as $\varepsilon \downarrow 0$. In particular, the noise does
   not depend on the spatial variable. Then, the dynamics of the phase-separating
   hyperplane Γ_t appearing in the limit is given by

   $$V = \kappa + c_0\dot{W}_t,$$

   where V is the inward normal velocity of Γ_t, κ is the mean curvature H of Γ_t
   multiplied by $d - 1$, and $c_0 = \frac{\sqrt{2}}{\int_{-1}^{1} \sqrt{V(u)}du}$.

Before starting the discussions for the stochastic case, we briefly summarize the
known results in a deterministic case, that is, for the PDE (4.1) with $\dot{W}^\varepsilon = 0$:

$$\partial_t u = \Delta u + \frac{1}{\varepsilon^2}f(u), \quad t > 0, \ x \in D. \tag{4.3}$$

Set $A(f) := \int_{-1}^{1} f(u)\, du = V(-1) - V(1)$, which is the difference of levels of the
two bottoms of the potential V corresponding to f such that $f = -V'$, recall Fig. 3.1.

A *traveling wave solution* $m = m(y)$, $y \in \mathbb{R}$, with speed $c = c(f) \in \mathbb{R}$ is determined as an increasing solution of the ordinary differential equation

$$\begin{cases} m'' + c\,m' + f(m) = 0, & y \in \mathbb{R}, \\ m(\pm\infty) = \pm 1, \end{cases}$$

where $m' = dm/dy$ and $m'' = d^2m/dy^2$. In particular, $v(t, y) = m(y - ct)$ is a solution of the one-dimensional reaction-diffusion equation

$$\partial_t v = \partial_y^2 v + f(v), \quad t > 0, \ y \in \mathbb{R}, \tag{4.4}$$

see [15, 82, 136]. We normalize the function m as $m(0) = 0$. It is known that $A(f)$ and $-c(f)$ have the same sign; especially, $A(f) = 0$ is equivalent to $c(f) = 0$. If $A(f) > 0$, then, since $V(-1) > V(1)$, the solution v moves from the metastable state -1 to the lower bottom $+1$ of V (i.e., a wave from -1 to $+1$ is created) and this yields a wave front m moving to the left so that $c(f) < 0$. We call "$A(f) = 0$" the balance condition.

We now come back to the equation (4.3). If $A(f) \neq 0$, the proper time scale is $O(\varepsilon)$, i.e., for the solution u^ε of (4.3), we have that

$$\bar{u}^\varepsilon(t, x) := u^\varepsilon(\varepsilon t, x) \longrightarrow \chi_{\Gamma_t}(x) \qquad (\varepsilon \downarrow 0),$$

where Γ_t is a hyperplane in D and $\chi_{\Gamma_t}(x) = 1$ (if $x \in$ outside of Γ_t), $\chi_{\Gamma_t}(x) = -1$ (if $x \in$ inside of Γ_t). The interface Γ_t evolves according to the *Huygens' principle*: waves with speed $c(f)$ are created at each point of Γ_t in all outward directions, and Γ_t is determined as the envelope of the wave fronts. See [69, 83, 84] for f of KPP type, Gärtner [120] for f of bistable type in higher-dimensional spaces, and [81] for f of bistable type in the one-dimensional space.

If $A(f) = 0$ and $d \geq 2$, we have that $c(f) = 0$ and therefore the wave front is immobile. Such a wave is called a *standing wave*. The above result shows that Γ_t does not move for the time scale $O(\varepsilon)$, so that one should consider a longer time scale. In fact, the proper time scale is $O(1)$, i.e., for the solution u^ε of (4.3),

$$u^\varepsilon(t, x) \longrightarrow \chi_{\Gamma_t}(x) \qquad (\varepsilon \downarrow 0)$$

and Γ_t evolves according to *motion by mean curvature*, see [73, 186] and many other references cited in Sect. 4.3.1. A heuristic argument to derive the motion by mean curvature is given in Sect. 4.2.2 with an additional noise.

If $A(f) = 0$ and $d = 1$, we can regard the one-dimensional wave front as a planar wave, so that the curvature is 0 (or we may think $\kappa = 0$ since $d - 1 = 0$) The proper time scale is therefore much longer than $O(1)$. In fact, Carr and Pego [44] showed that the proper scale is $O(\exp(C\varepsilon^{-1}))$, which is extremely long.

4.2 Sharp Interface Limits in a Stochastic Case

We consider the stochastic Allen-Cahn equation (4.1) only under the balance condition, so that our assumptions in this section are: $f \in C^\infty(\mathbb{R})$ is bistable satisfying the condition (4.2), $A(f) = 0$, together with the technical conditions

$$|f(u)| \le C(1 + |u|^p), \quad \sup_{u \in \mathbb{R}} f'(u) < \infty, \tag{4.5}$$

with some $C, p > 0$.

4.2.1 One-Dimensional Case with Space-Time White Noise

Let $\gamma > 0$ and $a \in C_0^2(\mathbb{R})$ be the intensity of the noise (we assume that it has a compact support to localize the problem and to kill the fluctuations near $x = \pm\infty$), and let $\dot{W}(t, x)$ be the space-time Gaussian white noise with the covariance structure (3.2). Under the conditions (4.5) on f, the SPDE (4.1) with $\dot{W}^\varepsilon(t, x) = \varepsilon^\gamma a(x) \dot{W}(t, x)$ has a unique solution (in the sense of generalized functions or in the mild sense) which is Hölder continuous:

$$u^\varepsilon(t, x) \in C^{\frac{1}{4} - , \frac{1}{2} -}((0, \infty) \times \mathbb{R}), \quad \text{a.s.}$$

Theorem 4.1 (Funaki [95, 96]). *If the initial value has the form $u^\varepsilon(0, x) = m((x - \xi)/\varepsilon)$ with some $\xi \in \mathbb{R}$ and the reaction term has the symmetry property $f(u) = -f(-u)$, then for all $\gamma > \frac{19}{2}$, we have the convergence in law*

$$\bar{u}^\varepsilon(t, x) := u^\varepsilon(\varepsilon^{-2\gamma - 1}t, x) \Longrightarrow \chi_{\xi_t}(x) \quad (\varepsilon \downarrow 0), \tag{4.6}$$

where $\chi_\xi(x) = 1_{\{x > \xi\}}$, $\chi_\xi(x) = -1_{\{x < \xi\}}$. The phase separation point ξ_t moves according to the following stochastic differential equation (SDE):

$$d\xi_t = \alpha_1 a(\xi_t) dB_t + \alpha_2 a(\xi_t) a'(\xi_t) dt, \quad \xi_0 = \xi, \tag{4.7}$$

where B_t is a one-dimensional Brownian motion, $\alpha_1 = \|m'\|_{L^2(\mathbb{R})}^{-1}$,

$$\alpha_2 = -\|m'\|_{L^2(\mathbb{R})}^{-2} \int_0^\infty dt \int_{\mathbb{R}^2} xp(t, x, y)^2 f''(m(y)) m'(y) dx dy,$$

and $p(t, x, y)$ is the fundamental solution of the linearized operator $\partial_t - \{\partial_y^2 + f'(m(y))\}$.

Physics behind and role of γ: Theorem 4.1 shows that the diffusion coefficient (mobility) α_1^2 is given by the inverse of the surface tension $\|m'\|_{L^2(\mathbb{R})}^2$ and this coincides with the conjecture made by Kawasaki and Ohta [162] and Spohn [205]. The condition that γ is sufficiently large guarantees that the effect of the reaction term $\frac{1}{\varepsilon^2}f$ dominates that of the random fluctuation term $\varepsilon^\gamma a(x)\dot{W}(t,x)$. If the random fluctuation is strong, that is, if $\gamma > 0$ is small, it might happen that the shape of the wave front (located at ξ_t) arising in the solution u is totally destroyed by the fluctuation. Note that the global minima of the Ginzburg-Landau-Wilson free energy $H(u)$ (with V replaced by $\frac{1}{\varepsilon^2}V$) given in (3.4) are $u \equiv +1$ and $u \equiv -1$ if V has bottoms with the same level at ± 1 (i.e., $A(f) = 0$) as in Fig. 3.1, and χ_ξ are merely local minima (called instantons). This suggests the possibility that due to the strong random fluctuation the interface disappears and the whole region is occupied by $+1$ or -1.

Meaning of time change: The time change $\varepsilon^{-2\gamma-1}$ in (4.6) is very different from the case without noise, recall the last part of Sect. 4.1. The intuitive reason for the properness of this time scale is as follows: $\bar{u} = \bar{u}^\varepsilon$ satisfies (in law) the SPDE

$$\partial_t \bar{u} = \varepsilon^{-2\gamma-1}\left\{\Delta\bar{u} + \frac{1}{\varepsilon^2}f(\bar{u})\right\} + \left(\varepsilon^{-2\gamma-1}\right)^{1/2} \cdot \varepsilon^\gamma a(x)\dot{W}(t,x). \tag{4.8}$$

Note that the noise term becomes $a(x)\varepsilon^{-1/2}\dot{W}(t,x)$. The strong drift $\varepsilon^{-2\gamma-1}$ pushes $\bar{u}$ to the neighborhood of

$$M^\varepsilon := \left\{\bar{u};\ \Delta\bar{u} + \frac{1}{\varepsilon^2}f(\bar{u}) = 0,\ \bar{u}(\pm\infty) = \pm 1\right\}$$

$$=\{m\left((x - \xi)/\varepsilon\right);\ \xi \in \mathbb{R}\},$$

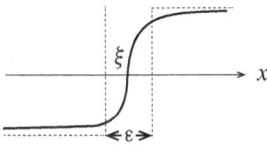

so that $\bar{u}^\varepsilon(t,x)$ is close to $m\left((x - \xi_t)/\varepsilon\right)$ with some ξ_t. In particular, the width of the interface is $O(\varepsilon)$. The contribution of the noise $\dot{W}(t,x)$ comes only from this region, therefore its order is $O(\varepsilon^{1/2})$ by self-similarity. This balances with the factor $\varepsilon^{-1/2}$ in front of the noise. On the other hand, since $\varepsilon^{-2\gamma-1} \ll \exp(C\varepsilon^{-1})$, the time scale is too short to observe the deterministic movement found by Carr and Pego [44].

Extensions of Theorem 4.1:

1. If the centering condition on f (i.e., the oddness of f) is violated (the balance condition $A(f) = 0$ is always assumed), we can show the law of large numbers

$$u^\varepsilon(\varepsilon^{-2\gamma}t,x) \Longrightarrow \chi_{\xi_t}(x), \qquad \dot{\xi}_t = \alpha_3 a^2(\xi_t), \tag{4.9}$$

with the constant

$$\alpha_3 = -\frac{1}{2\|m'\|^2_{L^2(\mathbb{R})}} \int_0^\infty dt \int_{\mathbb{R}^2} p(t, x, y)^2 f''(m(y))m'(y)dxdy.$$

The centering condition implies $\alpha_3 = 0$, so that we get the central limit theorem and obtain the random motion of the interface in the limit under the time scale longer than (4.9) as in Theorem 4.1. Brassesco, De Masi, and Presutti [40] discussed a similar problem at microscopic level under the centering condition. Brassesco and Buttà [39] studied the existence of non-odd function f for which $\alpha_3 \neq 0$ holds. See also Bertini, Brassesco, and Buttà [22]. The sharp interface limit for the invariant measures was studied by H. Weber [215].

2. S. Weber [217] studied the case where several interfaces (multi-kinks) appear on $[0, 1]$ under the periodic boundary conditions. The noise is $\varepsilon^\gamma \dot{W}(t, x)$ (the space-time white noise). Annihilating Brownian motions are obtained in the limit. His method is the following: First, consider approximate slow manifold $\mathcal{M}$ and a coordinate system around $\mathcal{M}$ (PDE case: Carr and Pego [44], X. Chen [51]). Secondly, use the idea of the expansion in the stochastic case due to Antonopoulou, Blömker, and Karali [13] when u is close to $\mathcal{M}$. Finally, show the annihilation when two interfaces touch. A related heuristic argument for deriving the stochastic multi-kink motion was given by Fatkullin, Kovačič, and Vanden Eijnden [77]. The stochastic multi-kink motion with coagulation can be derived from microscopic interacting Brownian particles in the zero temperature limit, see Funaki [102].

3. We can also study the self-similar space-time Gaussian (colored) noise $\{\dot{W}_h(t, x), 1/2 \leq h \leq 1\}$ with mean 0 and the covariance structure

$$E[\dot{W}_h(t, x)\dot{W}_h(s, y)] = \delta_0(t - s)Q_h(x - y),$$

where Q_h is the Riesz potential kernel of $(2h - 1)$th order:

$$Q_h(x) = \begin{cases} h(2h - 1)|x|^{2h-2}, & \text{if } 1/2 < h \leq 1, \\ \delta_0(x), & \text{if } h = 1/2. \end{cases}$$

Note that $\dot{W}_{\frac{1}{2}}(t, x)$ is the space-time Gaussian white noise and $\dot{W}_1(t, x) = \dot{W}(t)$ is the white noise in t. One can show that $\bar{u}^\varepsilon(t, x) := u^\varepsilon(\varepsilon^{-2\gamma-2h}t, x)$ converges to $\chi_{\xi_t}(x)$ as $\varepsilon \downarrow 0$ and the same SDE (4.7) is obtained in the limit with the constants α_1, α_2 modified according to the kernel Q_h, see Funaki [97]. It is reasonable to choose the self-similar noises from the viewpoint of the scalings.

Outline of the proof of Theorem 4.1: The proof consists of the following two steps:

1. To show that $\bar{u}^\varepsilon$ stays near M^ε, we take Ginzburg-Landau-Wilson free energy

$$H^\varepsilon(u) := \int_{\mathbb{R}} \left\{ \frac{1}{2} |\nabla u|^2(x) + \frac{1}{\varepsilon^2} V(u(x)) \right\} dx$$

as a Lyapunov function, where V is the potential corresponding to f (i.e., $f = -V'$). However, since u^ε is not differentiable in x, we cannot insert u^ε into H^ε so an extra trick is needed. The rate of convergence of u^ε to M^ε is controlled by the spectral gap property of the Hesse operator (Schrödinger operator) $-\partial_y^2 + V''(m)$ of $H^1(u)$ (i.e., $\varepsilon = 1$) at $u = m$.

2. We introduce a nice coordinate in a tubular neighborhood of M^ε (or on M^1 under the spatial scaling $x = \varepsilon y$). Consider the PDE (4.4). If its initial datum v_0 is in an L^2-tubular neighborhood of M^1, the solution $v = v(t, y)$ converges to a certain $m_\zeta(y) := m(y - \zeta)$ in M^1 as $t \to \infty$. The limit ζ depends on the initial value v_0 so that we denote it by $\zeta = \zeta(v_0) \in \mathbb{R}$. This defines a nice coordinate in an L^2-tubular neighborhood of M^1. In fact, if we compute the time derivative of $\zeta(\bar{u}^\varepsilon(t, \varepsilon\cdot))$, the diverging factor vanishes, see (4.11) and after.

We outline the derivation of the SDE (4.7): We first introduce $v^\varepsilon(t, y) := \bar{u}^\varepsilon(t, \varepsilon y)$ by observing $\bar{u}^\varepsilon(t, x)$ under the microscopic spatial variable y. Then, by (4.8), v^ε satisfies the SPDE

$$\partial_t v = \varepsilon^{-2\gamma-3} \{\Delta v + f(v)\} + \varepsilon^{-1} a(\varepsilon y) \dot{W}(t, y), \tag{4.10}$$

in the law sense. The coordinate $\zeta(v) \in \mathbb{R}$ defined in the L^2-tubular neighborhood of M^1 introduced above enjoys the following properties. We denote its first and second Fréchet derivatives by $D\zeta(y, v)$ and $D^2\zeta(y_1, y_2, v)$, respectively. The shifted standing wave m is defined by $m_\eta(y) = m(y - \eta)$, $y \in \mathbb{R}$, for $\eta \in \mathbb{R}$.

Lemma 4.1. 1. *For every v in the neighborhood of M^1, it holds that*

$$\langle D\zeta(\cdot, v), \Delta v + f(v) \rangle_{L^2(\mathbb{R})} = 0.$$

2. *For every $\eta \in \mathbb{R}$, it holds that*

$$D\zeta(y, m_\eta) = -\frac{m'_\eta(y)}{\|m'\|^2_{L^2(\mathbb{R})}}.$$

3. *For every $\eta \in \mathbb{R}$, it holds that*

$$D^2\zeta(y, y, m_\eta) = -\frac{1}{\|m'\|^2_{L^2(\mathbb{R})}} \int_0^\infty dt \int_{\mathbb{R}} p(t, y, z; m_\eta)^2 f''(m_\eta(z)) m'_\eta(z) \, dz,$$

where $p(t, y, z; m_\eta)$ denotes the fundamental solution of $\partial_t - \{\partial_y^2 + f'(m_\eta(y))\}$.

Indeed, the first follows from the observation that $\zeta(v(t)) = $ const in t along the solution $v(t)$ of the PDE (4.4). We can let $t \downarrow 0$ in the identity

$$0 = \frac{d}{dt}\zeta(v(t)) = \langle D\zeta(\cdot, v(t)), \Delta v(t) + f(v(t))\rangle_{L^2(\mathbb{R})}, \quad t > 0.$$

Another coordinate $\eta(v) \in \mathbb{R}$ in an L^2-tubular neighborhood of M^1 is defined as the minimizer of dist $(v, M^1) = \min_{\eta \in \mathbb{R}} \|v - m_\eta\|_{L^2(\mathbb{R})}$ and this is called the Fermi coordinate. The first Fréchet derivatives of the two coordinates $\zeta(v)$ and $\eta(v)$ actually coincide at $v = m_\eta$, however the second derivatives are different. The second and third identities in Lemma 4.1 follow by some concrete computations and we omit them.

Define the macroscopic phase separation point of $v^\varepsilon(t)$ by $\xi_t^\varepsilon := \varepsilon\zeta(v^\varepsilon(t))$. Then, applying Itô's formula and using the SPDE (4.10), we obtain

$$d\xi_t^\varepsilon = \int_{\mathbb{R}} D\zeta(y, v^\varepsilon(t))a(\varepsilon y)W(dtdy)$$

$$+ \frac{1}{2}\varepsilon^{-1}\int_{\mathbb{R}} D^2\zeta(y, y, v^\varepsilon(t))a^2(\varepsilon y)dy\, dt. \qquad (4.11)$$

Note that the diverging factor (the first term in (4.10)) vanishes due to Lemma 4.1-1. The quadratic variation of the first term in (4.11) is given by

$$\int_{\mathbb{R}} D\zeta(y, v^\varepsilon(t))^2 a^2(\varepsilon y)dy\, dt.$$

We can assume that $v^\varepsilon(t)$ is close to $m_{\varepsilon^{-1}\xi_t}$ for some ξ_t, and thus, by Lemma 4.1-2, this integral is close to

$$a^2(\xi_t)\int_{\mathbb{R}} \frac{(m_{\varepsilon^{-1}\xi_t}(y))^2}{\|m'\|_{L^2(\mathbb{R})}^4}dy\, dt = a^2(\xi_t)\alpha_1^2\, dt,$$

which leads to the first term in the SDE (4.7). Note that $a^2(\varepsilon y)$ is close to $a^2(\xi_t)$ under the integration in y as we explain below.

On the other hand, in the second term of formula (4.11), the contribution of $D^2\zeta(y, y, m_{\varepsilon^{-1}\xi_t})$ comes only from the vicinity of $y = \varepsilon^{-1}\xi_t$. Therefore, we can expand $a^2(\varepsilon y)$ as

$$a^2(\varepsilon y) = a^2(\xi_t) + \frac{1}{2}(a^2)'(\xi_t) \cdot \varepsilon(y - \varepsilon^{-1}\xi_t) + \cdots.$$

However, the contribution of the first term $a^2(\xi_t)$ vanishes upon integration with respect to y, since

$$\int_{\mathbb{R}} D^2\zeta(y, y, m_\eta)\, dy = 0,$$

which follows from Lemma 4.1-3 noting the symmetry of f. The contribution of the second term, after cancellation of ε^{-1} and ε, gives

$$\frac{1}{2}(a^2)'(\xi_t)\alpha_2\,dt,$$

from Lemma 4.1-3, and this is just the second term in the SDE (4.7).

4.2.2   Higher-Dimensional Case with Noise Asymptotically White in Time

In higher dimensions, as we pointed out, the stochastic Allen-Cahn equation with space-time white noise is ill-posed. For this reason, the sharp interface limit is discussed only for restricted types of noises, actually only for time-dependent noises which are independent of the spatial variables by now, except in Lions and Souganidis [177, 178].

Consider the stochastic Allen-Cahn equation (4.1) in higher dimensions with Neumann boundary condition: $\partial u/\partial n = 0$ $(x \in \partial D)$. We assume $A(f) = 0$, but don't require the symmetry (oddness) assumption on f. The noise $\dot{W}^\varepsilon(t, x) = \Xi_t^\varepsilon/\varepsilon$ depends only on t, and Ξ_t^ε has the form $\Xi_t^\varepsilon = \varepsilon^{-\gamma}\Xi(\varepsilon^{-2\gamma}t)$, $0 < \gamma < 2/3$, where $\Xi(t) \in C^1(\mathbb{R}_+)$ (a.s.) is a stationary process with mean 0 satisfying the strong mixing property. We have the convergence in law: $\Xi_t^\varepsilon \Rightarrow \alpha\dot{W}(t)$ $(\varepsilon \downarrow 0)$. Unfortunately, we cannot treat the white noise $\alpha\dot{W}(t)$ directly. Instead, we consider a mild noise Ξ_t^ε converging to $\alpha\dot{W}(t)$. Here, $W(t)$ is a one-dimensional Brownian motion and α is a constant given by

$$\alpha = \sqrt{2\int_0^\infty E[\Xi(0)\Xi(t)]\,dt}.$$

Then, under a certain condition on the initial value, which will be stated below in the outline of the proof, we have the following result for the solution $u^\varepsilon(t, x)$ of (4.1) with $\dot{W}^\varepsilon(t, x) = \Xi_t^\varepsilon/\varepsilon$ when $d = 2$ and $D \subset \mathbb{R}^2$:

$$\begin{cases} \partial_t u = \Delta u + \dfrac{1}{\varepsilon^2}f(u) + \dfrac{1}{\varepsilon}\Xi_t^\varepsilon, & t > 0,\ x \in D, \\ \partial u/\partial n = 0, & t > 0,\ x \in \partial D. \end{cases} \qquad (4.12)$$

Theorem 4.2 (Funaki [98]).   *As long as the limit phase separation curve Γ_t is strictly convex and stays inside D, we have the convergence in law:*

$$u^\varepsilon(t, x) \Longrightarrow \chi_{\Gamma_t}(x) \quad (\varepsilon \downarrow 0),$$

where the curve Γ_t moves according to the stochastically perturbed motion by curvature:

$$V = \kappa + (c_0\alpha)\dot{W}(t), \qquad (4.13)$$

where V denotes the inward normal velocity of Γ_t, κ is the curvature of Γ_t, and

$$c_0 = \frac{\sqrt{2}}{\int_{-1}^{1} du \sqrt{\int_u^1 f(v)\,dv}}.$$

H. Weber [216] extended Theorem 4.2 to arbitrary dimensions $d \geq 2$ and established the short-time sharp interface limit without assuming that the interfaces are convex. The convergence was shown in a.s.-sense due to the result by Dirr, Luckhaus, and Novaga [64], who gave pathwise solution to $V = \kappa + \dot{W}(t)$.

Heuristic derivation of (4.13): Since Ξ_t^ε is close to $\alpha\dot{W}(t)$, (4.1) is almost

$$\partial_t u = \Delta u + \frac{1}{\varepsilon^2}\{f(u) + \varepsilon\alpha\dot{W}(t)\}. \qquad (4.14)$$

In other words, the potential V is randomly perturbed to $V(u) - (\varepsilon\alpha\dot{W}(t))u$ and this yields a small traveling wave moving toward the minimizer of the perturbed fluctuating potential, see Fig. 4.2. This gives $c_0\alpha\dot{W}_t$ in (4.13). More precisely, for $a \in \mathbb{R}$ (with small $|a|$), define the traveling wave solution $m = m(y;a)$ and its speed $c = c(a)$ by

$$\begin{cases} m'' + cm' + \{f(m) + a\} = 0, & y \in \mathbb{R}, \\ m(\pm\infty) = m_\pm^*, \end{cases} \qquad (4.15)$$

where $m_\pm^* \equiv m_\pm^*(a) = \pm 1 + O(a)$ $(a \to 0)$ are solutions of $f(m_\pm^*) + a = 0$. Then, since the solution of (4.1) behaves as

$$u^\varepsilon(t,x) \sim m(d(x,\Gamma_t)/\varepsilon; \varepsilon\alpha\dot{W}(t)),$$

$$d(x,\Gamma_t) = \text{signed distance between } x \text{ and } \Gamma_t,$$

Fig. 4.2 Fluctuating potential

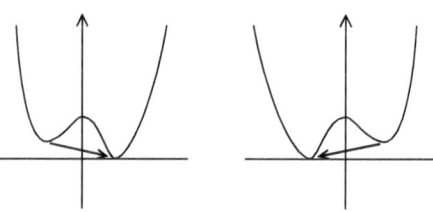

by inserting this in (4.1) or (4.14), we obtain

$$0 = \partial_t u^\varepsilon - \Delta u^\varepsilon - \frac{1}{\varepsilon^2} f(u^\varepsilon) - \frac{\alpha}{\varepsilon} \dot{W}(t)$$

$$\sim \frac{1}{\varepsilon} m'\left(\frac{d}{\varepsilon}\right) \partial_t d - \left\{ \frac{1}{\varepsilon} m'\left(\frac{d}{\varepsilon}\right) \Delta d + \frac{1}{\varepsilon^2} m''\left(\frac{d}{\varepsilon}\right) |\nabla d|^2 \right\}$$

$$- \frac{1}{\varepsilon^2} f(m) - \frac{\alpha}{\varepsilon} \dot{W}(t)$$

$$\sim \frac{1}{\varepsilon} m'\left(\frac{d}{\varepsilon}\right) \{ \partial_t d - \Delta d - c_0 \alpha \dot{W}(t) \}.$$

The last line follows from (4.15) and the facts that $|\nabla d| = 1$ near Γ_t and $c(a) = c(0) + c'(0)a + O(a^2) = -c_0 a + O(a^2)$. The ODE (4.15) was used to cancel the terms of order $O(1/\varepsilon^2)$. Thus the cancellation condition for the terms of the order $O(1/\varepsilon)$ becomes

$$\partial_t d = \Delta d + c_0 \alpha \dot{W}(t).$$

Since Δd describes the curvature on Γ_t, we obtain the limit equation (4.13).

Outline of the proof of Theorem 4.2: Since we assume the noise is mild, we can directly apply PDE methods, in particular, we can construct super/sub solutions of (4.1) using the comparison theorem. Those are given as functions close to $\tilde{u}^\varepsilon(t, x) := m(d(x, \Gamma_t^\varepsilon)/\varepsilon; \varepsilon \Xi_t^\varepsilon)$ (we assume this for $t = 0$), where the curve Γ_t^ε in D is determined by

$$V = \kappa - \frac{1}{\varepsilon} c(\varepsilon \Xi_t^\varepsilon). \tag{4.16}$$

However, if Γ_t^ε is convex, then in terms of the Gauss map $(\theta \in S^1 \simeq [0, 2\pi) \mapsto x(\theta) \in \Gamma_t^\varepsilon)$, (4.16) can be rewritten as a PDE for the curvature function $\kappa = \kappa^\varepsilon(t, \theta)$ $(:= $ curvature of Γ_t^ε at $x(\theta) \equiv x_t(\theta)) > 0$:

$$\partial_t \kappa = \kappa^2 \left\{ \partial_\theta^2 \kappa + \kappa - \frac{1}{\varepsilon} c(\varepsilon \Xi_t^\varepsilon) \right\}.$$

Then, one can study its limit as $\varepsilon \downarrow 0$ and obtain the following SPDE:

$$\partial_t \kappa = \kappa^2 \left\{ \partial_\theta^2 \kappa + \kappa + c_0 \alpha \circ \dot{W}(t) \right\}, \tag{4.17}$$

where $\circ$ denotes the Stratonovich stochastic integral. The equation (4.17) gives the precise mathematical meaning to the random perturbation of the curvature flow (4.13) in the convex setting.

Mass conserving stochastic Allen-Cahn equation: The Cahn-Hilliard equation and its stochastic version, that is the TDGL equation of conservative type, have a mass conservation law, as we saw in Sect. 3.1. Another equation that has such a property is the mass conserving Allen-Cahn equation, that is (4.18) with $\alpha = 0$. We add a stochastic term similarly to the above.

Let $u = u^\varepsilon(t, x)$ be the solution of the following SPDE in a smooth bounded domain D in $\mathbb{R}^d$:

$$\partial_t u^\varepsilon = \Delta u^\varepsilon + \frac{1}{\varepsilon^2}\left(f(u^\varepsilon) - \fint_D f(u^\varepsilon) \right) + \alpha \dot{W}^\varepsilon(t), \quad x \in D, \tag{4.18}$$

with Neumann boundary condition $\partial u^\varepsilon / \partial n = 0$ $(x \in \partial D)$, where $\alpha > 0$ and

$$\fint_D f(u^\varepsilon) = \frac{1}{|D|} \int_D f(u^\varepsilon(t, x)) dx.$$

Here, $\dot{W}^\varepsilon(t)$ is the time derivative of $W^\varepsilon(t) \in C([0, \infty))$ (a.s.) and $W^\varepsilon(t)$ converges to the one-dimensional Brownian motion $W(t)$ in a suitable sense. For example, we can take $W^\varepsilon(t) = \frac{1}{\alpha} \int_0^t \Xi_s^\varepsilon ds$ with Ξ_s^ε introduced above. The reaction term $f \in C^\infty(\mathbb{R})$ is bistable satisfying $A(f) = 0$. The mass conservation law is destroyed by the noise, but the following identity holds:

$$\fint_D u^\varepsilon(t) = \fint_D u^\varepsilon(0) + \alpha W^\varepsilon(t).$$

The noise term behaves as $O(\frac{1}{\varepsilon})$ in (4.12), while it is $O(1)$ in (4.18). This is a large difference in scaling between these two equations. Heuristically, one can expect $u^\varepsilon \approx \pm 1$ and this implies $\fint_D f(u^\varepsilon) = O(\varepsilon)$, so that $\frac{1}{\varepsilon^2} \fint_D f(u^\varepsilon)$ behaves as $O(\frac{1}{\varepsilon})$ and the evolution of this term is actually controlled by the noise, see Funaki and Yokoyama [118] for further discussions.

At least when $d = 2$ and phase separating curves remain convex, under suitable assumptions on the initial value u_0^ε, one can show that $u^\varepsilon(t, \cdot)$ converges to $\chi_{\Gamma_t}(\cdot)$ as $\varepsilon \downarrow 0$ locally in time. The evolution of limit curves $\Gamma_t \subset D$ is governed by

$$V = \kappa - \fint_{\Gamma_t} \kappa + \frac{\alpha|D|}{2|\Gamma_t|} \circ \dot{W}(t), \quad t \in [0, \sigma], \tag{4.19}$$

up to a certain stopping time $\sigma > 0$ (a.s.), where V is the inward normal velocity of Γ_t, κ is the (mean) curvature of Γ_t (multiplied by $d - 1$), and $\circ$ means Stratonovich stochastic integral, see [118].

The proof relies on the asymptotic expansion employed by Chen, Hilhorst, and Logak [52] in the case without noise. In this expansion, diverging terms like $(\dot{W}^\varepsilon(t))^2$, $(\dot{W}^\varepsilon(t))^3$ etc. appear. Usually, we cannot control such terms, but fortunately they appear only as higher order terms in the expansion. Therefore, if the diverging speed of the derivatives of $W^\varepsilon(t)$ is sufficiently small, we can control them.

4.3   A Brief Survey and Supplements

This section gives a brief survey on motions by mean curvature (MMC) with or without stochastic terms and the sharp interface limits leading to them. We also mention some supplementary results. Yip [223] is a nice review paper.

4.3.1   Deterministic Case

The MMC is the time evolution of $(d - 1)$-dimensional hypersurface Γ_t in $\mathbb{R}^d$ defined by

$$V = \kappa,$$

where V is an inward normal velocity and κ is the mean curvature multiplied by $d - 1$. The sign of κ (i.e.,the direction being inward or outward) is taken such that, for example, the sphere with radius $R = R(t)$ shrinks: $\dot{R} = -\frac{1}{R}$ or equivalently $\dot{\kappa} = \kappa^3$ for $\kappa = 1/R$. The MMC is a (negative) gradient flow for the surface area $\mathscr{A}(\Gamma) = |\Gamma|$.

MMC by classical methods: For the local existence of classical solutions, see Gage and Hamilton [119], Almgren, Taylor, and Wang [7], Evans and Spruck [75], who considered a PDE for signed distance functions.

For the global existence and long-time behaviors, see Gage and Hamilton [119], Grayson [129] for curves in two-dimensional space, Huisken [150] for convex surfaces, and Ecker and Huisken [67] for surfaces given by a graph. If Γ_t in $\mathbb{R}^d$ is represented as a graph $y = f(t, x), x \in \mathbb{R}^{d-1}, f$ evolves according to the PDE

$$\frac{\partial f}{\partial t} = \sqrt{1 + |\nabla f|^2} \, \text{div} \left(\frac{\nabla f}{\sqrt{1 + |\nabla f|^2}} \right).$$

Methods to handle solutions with singularities and topological changes: In general, classical solutions exist only locally in time, since singularities or topological changes may occur. Singularities in the case of self-similar structures are handled by Huisken [152]. A survey is given by Angenent [12].

For construction of solutions with singularities and topological changes, see Taylor, Cahn, and Handwerker [208], Soner [202, 203].

The existence of a global solution was first obtained by Brakke [38] in the setting of geometric measure theory and varifolds (this works only for isometric mean-curvature flow and solutions are not unique). Another approach, based on the level set formulation and viscosity solutions, was established by Evans and Spruck [74], Chen, Giga, and Goto [53] (solutions are global, unique, but fattening may happen). If Γ_t can be represented as $\Gamma_t = \{x \in \mathbb{R}^d; u(t, x) = 0\}$ with some $u(t, x)$, then $u(t, x)$ is a solution of the PDE

$$\frac{\partial u}{\partial t} = |\nabla u| \, \text{div} \left(\frac{\nabla u}{|\nabla u|} \right).$$

See the book of Giga [122] for this approach.

Sharp interface limit: Here we discuss the derivation of the MMC from the Allen-Cahn equation (4.3) with f satisfying the balance condition $A(f) = 0$.

de Mottoni and Schatzman [186] and X. Chen [49] applied the method of asymptotic expansions. These results were extended in the viscosity setting by Evans, Soner, and Souganidis [73], Soner [202, 203], Barles and Souganidis [18] and in varifold setting by Ilmanen [153]. See the book of Bellettini [19] for the sharp interface limit and the approach to the MMC by the equation for signed distances from the interfaces. See Alfaro, Hilhorst, and Matano [3] for the generation and the motion of interfaces under rather general initial data and Alfaro and Matano [4] for showing that the principal profile has actually the shape of the traveling wave solution m.

In one-dimension, X. Chen [51] classified time scales for the Allen-Cahn equation into four different stages. The third stage is Carr and Pego's super-slow motion. A numerical analytic study was carried out by Feng and Prohl [78].

Anisotropic MMC: Let a direction-dependent surface tension $\sigma = \sigma(\vec{n})$: $S^{d-1} \to \mathbb{R}_+$ be given, where $\vec{n}$ is an outward normal vector, and define the total surface tension of a hypersurface Γ by

$$J(\Gamma) := \int_{\Gamma} \sigma(\vec{n}(x)) dx,$$

where dx is the surface element of Γ, recall Sect. 1.1.1. Then, the σ-mean curvature κ_σ is defined as

$$\frac{d}{d\delta} J(\Gamma_\delta^g) \bigg|_{\delta=0} = \int_{\Gamma} \kappa_\sigma(\vec{n}(x)) \, g \cdot \vec{n}(x) dx,$$

where $\Gamma_\delta^g := \{x + \delta g(x); \, x \in \Gamma\}$ is a small perturbation of Γ in the direction of a vector field $g : \mathbb{R}^d \to \mathbb{R}^d$ taken as a test function. The anisotropic MMC (motion by σ-mean curvature) is defined by

$$V = \kappa_\sigma.$$

This is the (negative) gradient flow associated with J:

$$\frac{d}{dt} \Gamma_t = -\frac{\delta J}{\delta \Gamma} (\Gamma_t),$$

see Funaki and Spohn [116] for the derivation of such flow from the microscopic model.

The sharp interface limit leading to the anisotropic MMC is studied by Alfaro, Garcke, Hilhorst, Matano, and Schätzle [2]. Note that, if $\sigma \equiv 1$, $\kappa_\sigma = \kappa$ is the mean curvature.

Cahn-Hilliard equation and phase-field model: See Alikakos, Bates, and Chen [5], Chen [50], Cahn, Elliott, and Novick Cohen [41]. It is known that Caginalp's phase-field model converges to the Cahn-Hilliard equation as a certain parameter $\alpha \to 0$.

Volume preserving mean curvature flow: The time evolution of hypersurfaces Γ_t is governed by

$$V = \kappa - \mathrm{av}_{\Gamma_t}(\kappa),$$

where $\mathrm{av}_{\Gamma_t}(\kappa) = \frac{1}{|\Gamma_t|} \int_{\Gamma_t} \kappa$ is the average of κ over Γ_t, recall the last part of Sect. 4.2.2.

Huisken [151] showed the global existence of classical solutions when the initial hypersurface Γ_0 is uniformly convex. Elliott and Garcke [68] mostly studied the intermediate law, which converges to the volume-preserving mean curvature flow as the diffusion parameter $D \to \infty$, in the two-dimensional case. Escher and Simonett [72] established the local existence and uniqueness for general initial hypersurfaces, and the global existence and convergence to the sphere when the initial hypersurface is close to the sphere. The method is to rewrite the equation in a coordinate provided by a fixed reference hypersurface. (The demerit of this approach is that the noise term in the SPDE becomes multiplicative, which is more difficult to treat than the additive noise.)

4.3.2   Stochastic Case

Construction of the time evolution: Consider the stochastically perturbed motion by mean curvature (SMMC)

$$V = \kappa + \dot{W}(t, x),$$

where $W(t, x) = W^Q(t, x)$ is a Q-Brownian motion. This is an extension of (4.13), in which $W(t, x) = c_0 \alpha W(t)$.

If Γ_t is described by a graph $y = f(t, x)$, this can be rewritten as an SPDE for the height functions f:

$$\frac{\partial f}{\partial t} = \sqrt{1 + |\nabla f|^2}\ \mathrm{div}\left(\frac{\nabla f}{\sqrt{1 + |\nabla f|^2}}\right) + \sqrt{1 + |\nabla f|^2} \circ \dot{W}(t, x), \qquad (4.20)$$

which was derived by Kawasaki and Ohta [162, 163]. (In their case, the noise term is chosen to satisfy the fluctuation-dissipation relation.)

Lions and Souganidis [177, 178] give a meaning to this SPDE (4.20) or the evolution of Γ_t by introducing the notion of stochastic viscosity solutions for the level set function $u = u(t, x)$:

$$\frac{\partial u}{\partial t} = |\nabla u| \, \text{div} \left(\frac{\nabla u}{|\nabla u|} \right) + \alpha(x)|\nabla u| \circ \dot{W}(t),$$

when $W^Q(t, x) = \alpha(x)W(t)$ with smooth α. (This should admit a generalization to W^Q without much difficulty.) They also claim that the sharp interface limit shown by [98] can be proven in a more general setting and globally in time, i.e., past singularities, [178], Theorem 2.2.

Funaki [98] studied the SMMC by rewriting it as the SPDE (4.17) for curvature κ under the Gauss map. This approach has a limitation only for a convex setting in two-dimensional case. See Gurtin [135] and Giga and Mizoguchi [123] for the deterministic case.

The approach of Dirr, Luckhaus, and Novaga [64] followed by Weber [216] is different and relies on the analysis of signed distance functions $d(t, x) = $ signed dist(x, Γ_t), but is limited in an essential way to noises $\dot{W} = \dot{W}(t)$ dependent only on time to introduce a PDE for $q(t, x) := d(t, x) - W(t)$ to be dealt ω-wisely.

Von Renesse and coauthors [70, 71, 145] studied the problem when the interfaces are described as graphs. [71] treats a curve (represented as a graph) in $\mathbb{R}^2$; case (1) spatially-dependent (infinite-dimensional) noise, but for the vertical direction (i.e. y-axis-direction), SPDE (1.8); case (2) noise depending only on t for normal-to-curve direction, SPDE (1.10) = (4.20) with $\dot{W} = \dot{W}(t)$. [70] treats a special case of case (1) of [71] in which the vertical noise depends only on t and the x-variable. [145] establishes the existence of a solution in law sense for a surface (represented as a graph) in $\mathbb{R}^3$ with noise depending only on t, for the normal-to-surface direction, i.e., SPDE (4.20) with $\dot{W} = \dot{W}(t)$ and $x \in \mathbb{T}^2$.

Another evolution obtained by combining with a stochastic flow: Yip [222] considered the motion

$$V = \kappa + \langle \circ \dot{\mathbb{W}}(t, x), \mathbf{n} \rangle, \tag{4.21}$$

where $\mathbb{W}(t, x)$ is $\mathbb{R}^d$-valued (not $\mathbb{R}$-valued) and generates a stochastic flow $\{\varphi_{s,t}(x); 0 \leq s \leq t, x \in \mathbb{R}^d\}$ solving the SDE: $d\varphi_{s,t}(x) = \mathbb{W}(dt, \varphi_{s,t}(x)), t \geq s$, and $\varphi_{s,s}(x) = x$. This dynamics is different from those discussed above. He constructed a solution by means of a splitting method, patching the mean-curvature flow and the stochastic flow locally in time (similar to Funaki [94]).

Sharp interface limit: The derivation of the stochastically perturbed mean-curvature motion obtained in the sharp interface limit for the stochastic Allen-Cahn equation was initiated by physics papers by Kawasaki and Ohta. This was followed in one-dimension by mathematical papers, recall Sect. 4.2.1.

Katsoulakis, Kossioris, and Lakkis [160] studied the one-dimensional stochastic Allen-Cahn equation using numerical computation (discretization) and regularization of the space-time white noise.

Even in an ill-posed setting in higher-dimensions, it might be possible to apply the formulations of Hairer or Gubinelli noting the theory of Glimm, Jaffe, and Spencer [124–126] for the phase transition in the $P(\phi)_2$-model. (Remarks made by D. Bridges and G. Slade.)

Generation of interfaces in a stochastic setting was studied by Alfaro, Antonopoulou, Karali, and Matano [1] and Lee [175, 176].

Röger and Weber [196] tried to derive the motion (4.21) discussed by Yip in the sharp interface limit for a modified stochastic Allen-Cahn equation (with multiplicative noise):

$$\frac{\partial u}{\partial t} = \Delta u + \frac{1}{\varepsilon^2} f(u) + \nabla u \circ \dot{\mathbb{W}}(t, x).$$

They also used a splitting method to solve this SPDE and showed the tightness.

Stochastic Cahn-Hilliard equation: Antonopoulou, Blömker, and Karali [13] studied the Cahn-Hilliard equation with smooth noise (TDGL equation of conservative type with Q-Brownian motion in place of space-time white noise) on $[0, 1]$ with no-flux boundary conditions of Neumann type:

$$\partial_x u = \partial_x^3 u = 0 \quad \text{at } x = 0, 1.$$

SDEs are obtained in the sharp interface limit for multi-kinks before collisions (the result is local in time).

Bertini, Brassesco, Buttà, and Presutti [21] studied a one-dimensional phase field model with a colored noise, which has a conservation law. In a sharp interface limit, they derived a non-Markovian process for the front evolution in the limit.

Bertini, Brassesco, and Buttà [23] considered the Cahn-Hilliard equation (3.6) with the noise $\sqrt{-\Delta} \dot{W}(t, x)$ multiplied by ε. The motion of the phase separation point is infinitesimally governed by a fractional Brownian motion with self-similarity parameter $\frac{1}{4}$.

Antonopoulou, Karali, and Kossioris [14] considered the Cahn-Hilliard equation with deterministic noise (under white noise scaling) and gave a formal expansion of the solutions. For the deterministic Cahn-Hilliard equation, it is known that the Hele-Shaw free boundary problem appears in the sharp interface limit (instead of mean curvature motion in the Allen-Cahn case).

Role of stochasticity: Dirr, Luckhaus, and Novaga [64] and Souganidis and Yip [204] used the SMMC with $\dot{W} = \varepsilon \dot{W}(t)$ (only time-dependent noise) to select a right (unique) solution among the non-unique deterministic solutions as $\varepsilon \downarrow 0$. The philosophy is similar to the problems discussed in Chap. 1.

Bertini, Buttà, and Pisante [24] has recently proved the upper bound of the LDP for the stochastic Allen-Cahn equation with a small parameter in front of the noise, which is the space-time white noise smeared depending on ε. The minimizers of the corresponding rate function are Brakke solutions for the MMC, so that they are not unique. A further problem would be to select a proper solution from them by deriving more precise estimates than the LDP.

Chapter 5
KPZ Equation

The KPZ equation is a kind of stochastic PDE which describes the motion of a growing interface with random fluctuation, but known to be ill-posed and to require a renormalization. To give a meaning, we introduce two types of its approximations with diverging renormalization factors. One is simple, in which we replace the space-time white noise by its smeared noise and subtract a proper diverging constant. Then, one can see that the limit is the so-called Cole-Hopf solution of the KPZ equation. Another approximation is suitable for studying the invariant measures. Under the Cole-Hopf transformation, this approximating equation can be rewritten as a linear heat equation with an additional nonlinear term. We show by applying the Boltzmann-Gibbs principle that this nonlinear term can be asymptotically replaced by a linear term and find out that the solution of the approximating KPZ equation of second type converges to the Cole-Hopf solution with the term $\frac{t}{24}$ added, at least in the stationary situation. This result is generalized to multi-component coupled KPZ equation, which plays an important role in the theory of fluctuating hydrodynamics, due to the paracontrolled calculus.

5.1   The KPZ Equation, Its Ill-Posedness, and Its Renormalization

Kardar, Parisi, and Zhang [159] introduced the following SPDE for the height function $h(t, x)$ of a growing interface with random fluctuations:

$$\partial_t h = \frac{1}{2}\partial_x^2 h + \frac{1}{2}\left(\partial_x h\right)^2 + \dot{W}(t, x), \quad t > 0, \ x \in \mathbb{R} \ (\text{or } \mathbb{T} = \mathbb{R}/\mathbb{Z}), \tag{5.1}$$

where $\dot{W}(t, x)$ is the space-time Gaussian white noise with covariance structure (3.2), see Fig. 5.1. The coefficients $\frac{1}{2}$ are not important, since we can change them by space-time scaling.

© The Author(s) 2016
T. Funaki, *Lectures on Random Interfaces*, SpringerBriefs in Probability and Mathematical Statistics, DOI 10.1007/978-981-10-0849-8_5

Fig. 5.1 Color changes in time

The KPZ equation (5.1) is actually ill-posed. Indeed, in (5.1), dropping the nonlinear term, we have seen in Chap. 3 that $h \in C^{\frac{1}{4}-,\frac{1}{2}-}((0,\infty) \times \mathbb{R})$ a.s. and for such h the spatial derivative $\partial_x h$ is defined only in the sense of generalized functions. Therefore, the nonlinear term $(\partial_x h)^2$ is not definable in the usual sense. The roughness caused by the noise and the nonlinearity do not match. Instead, the following *renormalized KPZ equation* with compensator $\delta_x(x)$ $(= +\infty)$ has a meaning:

$$\partial_t h = \frac{1}{2}\partial_x^2 h + \frac{1}{2}\left\{(\partial_x h)^2 - \delta_x(x)\right\} + \dot{W}(t,x), \quad x \in \mathbb{R}, \tag{5.2}$$

as we will see later in Sect. 5.2 by applying the Cole-Hopf transformation.

Approaches using regularity structures, paracontrolled calculus, and probability energy solutions: Hairer [138, 139] and Gubinelli and Perkowski [131] discussed the renormalized KPZ equation (5.2) based on regularity structures and paracontrolled calculus, respectively, recall Sect. 3.1. In particular, their methods do not rely on the Cole-Hopf transformation. Hairer and Shen [144] proved that the solution of the KPZ equation driven by a stationary centered space-time random field with the mixing property converges to that of the KPZ equation (5.1), or, more precisely to its Cole-Hopf solution defined by (5.7) below. Hairer and Quastel [143] considered the equation $\partial_t h = \partial_x^2 h + \varepsilon F(\partial_x h) + \delta \eta$, where η is a smooth space-time Gaussian noise, and discussed the convergence of the solution to the Cole-Hopf solution under certain rescalings in the parameters ε and δ. See also Gubinelli and Perkowski [133] for a related result. Hoshino [148] discussed the KPZ equation with less regular noises $\partial_x^\gamma \dot{W}(t,x), 0 \le \gamma < \frac{1}{4}$, where ∂_x^γ denote fractional derivatives. For this equation, $\gamma = \frac{1}{2}$ is critical (see Sect. 3.1), but the theory is applicable up to $\gamma < \frac{1}{4}$.

Gonçalves and Jara [127], Gonçalves, Jara, and Sethuraman [128] introduced a notion of *probability energy solutions* to the KPZ equation or the corresponding stochastic Burgers equation (5.3). This is a formulation using a kind of martingale problems. The uniqueness of the stationary energy solutions was shown by Gubinelli and Perkowski [132].

Heuristic derivation of the KPZ equation: If a curve $\mathscr{C}_t = \{(x,y); y = h(t,x), x \in \mathbb{R}\}$ located in the plane $\mathbb{R}^2$ evolves upward with normal velocity $V = \kappa + A$, where κ is the (signed) curvature of $\mathscr{C}_t$ and $A > 0$ is a constant, then its height function $h(t,x)$ satisfies the nonlinear PDE

$$\partial_t h = \frac{\partial_x^2 h}{1 + (\partial_x h)^2} + A(1 + (\partial_x h)^2)^{1/2}.$$

Indeed, this is exactly the same equation as (4.20) with $\circ\dot{W}$ replaced by A in one-dimensional case, see also Matano, Nakamura, and Lou [180]. The KPZ equation (5.1) is obtained by taking the leading terms in this equation (more precisely, in the equation for $h(t, x) - At$ rather than $h(t, x)$ itself) under the assumption that the tilt $\partial_x h$ of the interface is small and taking the fluctuations caused by a space-time white noise into account. Note that $\frac{1}{2}$ is put in front of the second derivative and A is chosen as $A = 1$. This simplification is essential in view of the scaling property or universality of the KPZ equation.

Large-time behavior: The study of the KPZ equation has recently attracted a lot of attention. Balász, Quastel, and Seppäläinen [17] showed the $\frac{1}{3}$-power law in the stationary case, that is,

$$ct^{\frac{2}{3}} \le \text{Var}(h(t, 0)) \le Ct^{\frac{2}{3}},$$

holds, in other words, the fluctuations of $h(t, 0)$ are of order $t^{\frac{1}{3}}$. This is a very different behavior from the usual central limit theorem. The limit distributions of $h(t, 0)$ as $t \to \infty$ are also studied. For the narrow wedge initial distribution, Sasamoto and Spohn [199, 200] and Amir, Corwin, and Quastel [10] showed that the limit distribution of $h(t, 0)$ under the scaling is given by the so-called GUE-Tracy-Widom distribution; see also Quastel [195]. Borodin, Corwin, Ferrari, and Veto [34] studied the stationary case. These results were obtained based on the exactly solvable structure of the KPZ equation. The method is sometimes called "*Integrable probability*".

  Studies on models in KPZ universality class also should be mentioned briefly. TASEP has a determinantal structure and been studied well since Johansson [157]. ASEP and q-TASEP are also studied.

Derivation of KPZ equation from microscopic interacting systems: Bertini and Giacomin [27] derived the Cole-Hopf solution of the KPZ equation from the weakly ASEP. Corwin and Tsai [54], Corwin, Shen, and Tsai [55] derived it from the stochastic higher spin vertex model (which is an integrable system), higher spin ASEP (which has a duality), and ASEP(q, j). Gubinelli and Perkowski [131] derived it from the $\nabla\varphi$-interface model.

5.2   Cole-Hopf Solution and the Linear Stochastic Heat Equation

One simple approach to the renormalized KPZ equation (5.2) relies on the Cole-Hopf transformation. Consider the tilt (gradient field) of the interface $u(t, x) = \partial_x h(t, x)$, which satisfies the viscous stochastic Burgers equation from (5.1),

$$\partial_t u = \frac{1}{2}\partial_x^2 u + \frac{1}{2}\partial_x u^2 + \partial_x \dot{W}(t,x). \tag{5.3}$$

Here the noise has even less regularity. As we discussed in Chap. 2, the Cole-Hopf transformation (2.31) brings the viscous Burgers equation without noise, which arises as a hydrodynamic equation for the gradient field of the heights, to a linear heat equation. Motivated by this, we formally apply the Cole-Hopf transformation for the solution $h(t,x)$ of (5.1):

$$Z(t,x) := e^{-\int_x^\infty u(t,y)dy} = e^{h(t,x)-h(t,\infty)},$$

or

$$Z(t,x) := e^{h(t,x)}. \tag{5.4}$$

This leads to the linear SPDE with a multiplicative noise

$$\partial_t Z = \frac{1}{2}\partial_x^2 Z + Z\dot{W}_t(x), \tag{5.5}$$

where the last multiplicative term should be interpreted in Stratonovich's sense (later we interpret it in Itô's sense and call (5.5) SHE). Indeed, (5.5) is derived by noting that $\partial_t Z = Z\partial_t h$ under the Stratonovich calculus, inserting the right-hand side of (5.1) for $\partial_t h$, and finally noting that

$$\partial_x^2 Z = Z\{\partial_x^2 h + (\partial_x h)^2\}, \tag{5.6}$$

which follows from (5.4) by easy computation. Unfortunately, the SPDE (5.5) is ill-posed if we interpret the last term in Stratonovich's sense.

However, this SPDE turns out to be well-posed if we interpret the last term in Itô's sense. Namely, the solution $Z(t)$ of (5.5) in Itô's sense is defined in the sense of generalized functions or in a mild form:

$$Z(t,x) = \int_{\mathbb{R}} p(t,x,y)Z_0(y)dy + \int_0^t \int_{\mathbb{R}} p(t-s,x,y)Z(s,y)W(dsdy),$$

with the heat kernel $p(t,x,y) = \frac{1}{\sqrt{2\pi t}}e^{-(y-x)^2/(2t)}$, where the last term is to be understood as a stochastic integral in Itô's sense, recall Sect. 3.3.1. It is known that these two notions are equivalent, and there exists a unique solution satisfying $Z \in C([0,\infty) \times \mathbb{R})$ and $\sup_{x \in \mathbb{R}} e^{-r|x|}|Z(t,x)| < \infty$ for every $r > 0$ a.s. (if this holds at $t = 0$). Moreover, the following strong comparison theorem is known: If $Z(0,x) \geq 0$ for all $x \in \mathbb{R}$ and $Z(0,x) > 0$ for some $x \in \mathbb{R}$, then $Z(t,x) > 0$ holds for every $t > 0, x \in \mathbb{R}$ a.s., see Mueller [187], Shiga [201]. Therefore, one can define the (inverse) Cole-Hopf transformation

$$h(t,x) := \log Z(t,x). \tag{5.7}$$

Such h is meaningful and is called the *Cole-Hopf solution* of the KPZ equation; it was first introduced by Bertini and Giacomin [27], although the equation (5.1) does not make sense. The SPDE (5.5) with the multiplicative term interpreted in Itô's sense is called the *linear stochastic heat equation* (SHE).

Now, we give a heuristic derivation of the renormalized KPZ equation (5.2) from the (well-posed) SHE (5.5) under the Cole-Hopf transformation (5.7). Applying Itô's formula for $h = \log z$, we have that

$$
\begin{aligned}
\partial_t h &= Z^{-1}\partial_t Z - \frac{1}{2}Z^{-2}(\partial_t Z)^2 \\
&= Z^{-1}\left(\frac{1}{2}\partial_x^2 Z + Z\dot{W}\right) - \frac{1}{2}\delta_x(x) \\
&= \frac{1}{2}\{\partial_x^2 h + (\partial_x h)^2\} + \dot{W} - \frac{1}{2}\delta_x(x),
\end{aligned}
$$

where the second term in the first line is the Itô correction term, the second line follows by the SHE (5.5) noting $(dZ(t,x))^2 = (ZdW(t,x))^2$ and $dW(t,x)dW(t,y) = \delta(x-y)dt$, which in turn follows from the covariance structure (3.2) of the space-time white noise, and the third line follows from (5.6). This leads to the renormalized KPZ equation (5.2).

5.3 KPZ Approximating Equations

The argument in the last section was rather heuristic. In order to make it rigorous, let us introduce approximations for (5.2). One is well adapted to studying invariant measures. The results of Sects. 5.3.2, 5.4, and 5.5 are taken from Funaki and Quastel [112].

5.3.1 Simple Approximation

Let us first introduce a symmetric convolution kernel. Take $\eta \in C_0^\infty(\mathbb{R})$ satisfying $\eta(x) \geq 0$, $\eta(x) = \eta(-x)$ and $\int_{\mathbb{R}} \eta(x)dx = 1$, and set $\eta^\varepsilon(x) := \frac{1}{\varepsilon}\eta(\frac{x}{\varepsilon})$ for $\varepsilon > 0$. Then, we define the smeared noise by

$$
W^\varepsilon(t,x) = \langle W(t), \eta^\varepsilon(x - \cdot)\rangle \ \left(= W(t) * \eta^\varepsilon(x)\right).
$$

The first KPZ approximating equation is given by

$$
\partial_t h = \frac{1}{2}\partial_x^2 h + \frac{1}{2}\left((\partial_x h)^2 - c^\varepsilon\right) + \dot{W}^\varepsilon(t,x), \tag{5.8}
$$

where we choose the renormalization factor $c^\varepsilon = \eta_2^\varepsilon(0)$ $(:= \eta^\varepsilon * \eta^\varepsilon(0)) = \frac{1}{\varepsilon}\|\eta\|^2_{L^2(\mathbb{R})}$, which diverges as $\varepsilon \downarrow 0$. Then, by applying Itô's formula for $Z(t, x) := e^{h(t,x)}$, and noting that $(dh(t, x))^2 = (dW^\varepsilon(t, x))^2 = c^\varepsilon dt$ and (5.6), we obtain

$$\partial_t Z = \frac{1}{2}\partial_x^2 Z + Z\dot{W}^\varepsilon(t, x), \tag{5.9}$$

where the last multiplicative term is defined in Itô's sense. It was shown by Bertini and Giacomin [27] that $Z = Z^\varepsilon$ converges to the solution Z of the SHE (5.5), and therefore the solution $h = h^\varepsilon$ of the first approximating equation (5.8) converges to the Cole-Hopf solution $h_{CH}(t, x)$ of the KPZ equation defined by (5.7) as $\varepsilon \downarrow 0$.

5.3.2 Approximation Adapted to Studying Invariant Measures

We introduce another approximation which is suitable for studying the invariant measures of the KPZ equation. The general principle behind introducing this approximation is the following. Consider the SPDE

$$\partial_t h = F(h) + \dot{W},$$

where F is a nonlinear operator acting on h, and let A be some operator. In our case, $F(h)(x) = \frac{1}{2}\partial_x^2 h + \frac{1}{2}(\partial_x h)^2$. Then, the class of the invariant measures essentially does not change or even becomes richer due to the presence of A for the SPDE

$$\partial_t h = A^2 F(h) + A\dot{W}.$$

Recall that the TDGL equation (3.3) has this property with $A = (-\Delta)^{\alpha/2}$, or if A is just a constant, this being a simple result of the time change. This principle may not be true in a non-reversible situation, but we try to apply the idea in our setting.

We introduce the second KPZ approximating equation,

$$\partial_t h = \frac{1}{2}\partial_x^2 h + \frac{1}{2}\left((\partial_x h)^2 - c^\varepsilon\right) * \eta_2^\varepsilon + \dot{W}^\varepsilon(t, x), \tag{5.10}$$

where $\eta_2(x) = \eta * \eta(x)$, $\eta_2^\varepsilon(x) = \frac{1}{\varepsilon}\eta_2(\frac{x}{\varepsilon})$ and $c^\varepsilon = \eta_2^\varepsilon(0)$ as before. Since the operators ∂_x^2 and $*\eta_2^\varepsilon$ commute, we do not put $*\eta_2^\varepsilon$ in the first term of (5.10). This changes the invariant measure, as we will see in the following theorem. Note that the solution h of (5.10) is smooth in x, so that we can consider the associated tilt process $\partial_x h$.

Invariant measures at approximating level: Let ν^ε be the distribution of $\partial_x(B * \eta^\varepsilon(x))$, where $B = \{B(x)\}_{x \in \mathbb{R}}$ is the two-sided Brownian motion. The measure ν^ε is independent of the choice of $B(0)$, so we take $B(0) = 0$.

Theorem 5.1. *The probability measure ν^ε is invariant for the tilt process $\partial_x h$ determined by the SPDE (5.10).*

Da Prato, Debussche, and Tubaro [56] studied a similar SPDE to (5.10) on $\mathbb{T}$.

Outline of the proof of Theorem 5.1: The proof consists of the following three steps: First, we introduce the discretization of the SPDE (5.10) (on $\mathbb{T}$) on a discrete torus $\mathbb{T}^{(N)} = \{1, 2, \ldots, N\}$. The discretization of the nonlinear term $(\partial_x h)^2$ should be carefully chosen (cf. Sasamoto and Spohn [198]):

$$\frac{1}{3}\left\{(h_{i+1} - h_i)^2 + (h_i - h_{i-1})^2 + (h_{i+1} - h_i)(h_i - h_{i-1})\right\}, \quad i \in \mathbb{T}_N.$$

Then, one can show that the discrete version of ν^ε defined on $\mathbb{T}^{(N)}$ is invariant. Secondly, the continuum limit as $N \to \infty$ leads to the result on $\mathbb{T}$. This can be easily extended to a torus $\mathbb{T}_M = \mathbb{R}/M\mathbb{Z}$ of size M. Finally, we take an infinite-volume limit as $M \to \infty$ by the usual tightness and martingale problem approach. This proves the conclusion on $\mathbb{R}$.   □

Remark 5.1. The infinitesimal invariance can be directly shown based on the Wiener-Itô expansion of tame functions Φ, see [107]:

$$\int \mathscr{L}^\varepsilon \Phi(h)\nu^\varepsilon(dh) = 0, \tag{5.11}$$

where $\mathscr{L}^\varepsilon$ is the (pre) generator associated with the SPDE (5.10), given by

$$\mathscr{L}^\varepsilon = \mathscr{L}_0^\varepsilon + \mathscr{A}^\varepsilon,$$

$$\mathscr{L}_0^\varepsilon \Phi(h) = \frac{1}{2}\int_{\mathbb{R}^2} D^2\Phi(x_1, x_2; h)\eta_2^\varepsilon(x_1 - x_2)dx_1 dx_2 + \frac{1}{2}\int_{\mathbb{R}} \partial_x^2 h(x)D\Phi(x; h)dx,$$

$$\mathscr{A}^\varepsilon \Phi(h) = \frac{1}{2}\int_{\mathbb{R}} \left((\partial_x h)^2 - c^\varepsilon\right) * \eta_2^\varepsilon(x)D\Phi(x; h)dx,$$

and $D\Phi$ and $D^2\Phi$ are the Fréchet derivatives of Φ.

5.4 Passing to the Limit $\varepsilon \downarrow 0$

It was easy to pass to the limit $\varepsilon \downarrow 0$ for the first KPZ approximating equation (5.8). Our next goal is to pass to the limit $\varepsilon \downarrow 0$ for the second KPZ approximating equation (5.10). We consider its Cole-Hopf transformation $Z(\equiv Z^\varepsilon) := e^h$ as before. Then, by Itô's formula, Z satisfies the SPDE

$$\partial_t Z = \frac{1}{2}\partial_x^2 Z + A^\varepsilon(x, Z) + Z\dot{W}^\varepsilon(t, x), \tag{5.12}$$

where the last term is defined in Itô's sense and

$$A^{\varepsilon}(x, Z) = \frac{1}{2}Z(x)\left\{\left(\frac{\partial_x Z}{Z}\right)^2 * \eta_2^{\varepsilon}(x) - \left(\frac{\partial_x Z}{Z}\right)^2(x)\right\}.$$

In fact, this term was 0 as we saw in (5.9) for (5.8), but here it appears as the result of putting $*\eta_2^{\varepsilon}$ in (5.10). The term $A^{\varepsilon}(x, Z)$ looks like it would vanish as $\varepsilon \downarrow 0$ (at least if $Z = Z^{\varepsilon}$ behaves nicely). But this is not true.

Indeed, if we take the average in time t and space x, $A^{\varepsilon}(x, Z)$ can be replaced by a linear term $\frac{1}{24}Z$, as we will see in Theorem 5.2 and after. Thus, the limit equation of (5.12) as $\varepsilon \downarrow 0$ becomes (in the stationary case for the tilt process)

$$\partial_t Z = \frac{1}{2}\partial_x^2 Z + \frac{1}{24}Z + Z\dot{W}(t, x). \tag{5.13}$$

Or, heuristically at the KPZ level,

$$\partial_t h = \frac{1}{2}\partial_x^2 h + \frac{1}{2}\{(\partial_x h)^2 - \delta_x(x)\} + \frac{1}{24} + \dot{W}(t, x).$$

We can also say that the limit $h(t, x)$ of the solution $h^{\varepsilon}(t, x)$ of the second KPZ approximating equation (5.10) is given by

$$h(t, x) = h_{\mathrm{CH}}(t, x) + \frac{1}{24}t, \tag{5.14}$$

where $h_{\mathrm{CH}}(t, x)$ is the Cole-Hopf solution of the KPZ equation.

We apply a method similar to that used to show the so-called *Boltzmann-Gibbs principle* to prove that $A^{\varepsilon}(x, Z^{\varepsilon}(s))$ can be replaced asymptotically by $\frac{1}{24}Z^{\varepsilon}(s, x)$. To avoid the complexity arising from the infiniteness of the invariant measures, we consider $h^{\varepsilon}(t, \rho) := \int h^{\varepsilon}(t, x)\rho(x)dx$, the height averaged by $\rho \in C_0^{\infty}(\mathbb{R}), \geq 0, \int \rho(x)dx = 1$, in modulo 1 and call it the *wrapped process*.

Theorem 5.2. *For every test function $\varphi \in C_0(\mathbb{R})$ satisfying supp $\varphi \cap$ supp $\rho = \emptyset$, we have that*

$$\lim_{\varepsilon \downarrow 0} E^{\pi \otimes \nu^{\varepsilon}}\left[\left\{\int_0^t \tilde{A}^{\varepsilon}(\varphi, Z^{\varepsilon}(s))ds\right\}^2\right] = 0, \tag{5.15}$$

where π is the uniform probability measure on $[0, 1)$ for $h^{\varepsilon}(0, \rho) \in [0, 1)$,

$$\tilde{A}^{\varepsilon}(\varphi, Z) = \int_{\mathbb{R}} \tilde{A}^{\varepsilon}(x, Z)\varphi(x)dx,$$

and

$$\tilde{A}^{\varepsilon}(x, Z) = A^{\varepsilon}(x, Z) - \frac{1}{24}Z(x).$$

Proof. The first step is the reduction of the equilibrium dynamic problem (5.15) to a static one: The expectation in (5.15) is bounded by

$$20\, t \sup_{\Phi \in L^2(\pi \otimes \nu^{\varepsilon})} \left\{ 2E^{\pi \otimes \nu^{\varepsilon}} \left[\tilde{A}^{\varepsilon}(\varphi, Z)\Phi \right] - \langle \Phi, (-\mathscr{L}_0^{\varepsilon})\Phi \rangle_{\pi \otimes \nu^{\varepsilon}} \right\}, \qquad (5.16)$$

where $\mathscr{L}_0^{\varepsilon}$ is the symmetric part of $\mathscr{L}^{\varepsilon}$ stated in Remark 5.1. This is a generic bound that holds in a stationary situation, cf. Komorowski, Landim, and Olla [170]. Here, we can rewrite

$$2E^{\pi \otimes \nu^{\varepsilon}} \left[\tilde{A}^{\varepsilon}(\varphi, Z)\Phi \right] = E^{\pi} \left[Z_{\rho} E^{\nu^{\varepsilon}} \left[B^{\varepsilon}(\varphi, Z)\Phi(h(\rho), \nabla h) \right] \right],$$

where $Z_{\rho} = \exp\{\int_{\mathbb{R}} \log Z(x)\rho(x)dx\} \in [1, e)$, $B^{\varepsilon}(x, Z) = \frac{2A^{\varepsilon}(x, Z)}{Z_{\rho}}$ and $B^{\varepsilon}(\varphi, Z) = \int_{\mathbb{R}} B^{\varepsilon}(x, Z)\varphi(x)dx$. We call $\nabla h = \{h(x) - h(y); x, y \in \mathbb{R}\}$ the tilt of h and regard $\nabla h \in \tilde{\mathscr{C}} := \mathscr{C}/\sim$, where $\mathscr{C} = C(\mathbb{R})$ and $h \sim g$ if $h - g = $ constant for $h, g \in \mathscr{C}$. Note that $B^{\varepsilon}(x, Z)$ is a tilt variable.

Then, the key is the following static bound: ν on $\tilde{\mathscr{C}}$ is the distribution of $\{B(x) - B(0)\}_{x \in \mathbb{R}}$, where B is the two-sided Brownian motion.

Proposition 5.1. *Let $\Phi = \Phi(\nabla h) \in L^2(\tilde{\mathscr{C}}, \nu)$ satisfy $\|\Phi\|_{1,\varepsilon}^2 = \langle \Phi, (-\mathscr{L}_0^{\varepsilon})\Phi \rangle_{\pi \otimes \nu^{\varepsilon}} < \infty$, and φ satisfy the condition of Theorem 5.2. Then*

$$\left| E^{\nu^{\varepsilon}} \left[B^{\varepsilon}(\varphi, Z)\Phi \right] \right| \le C(\varphi)\sqrt{\varepsilon}\|\Phi\|_{1,\varepsilon}, \qquad (5.17)$$

with some positive constant $C(\varphi)$, which depends only on φ, for all sufficiently small $\varepsilon > 0$.

Once this proposition is shown, the proof of Theorem 5.2 is concluded, since the sup in (5.16) is bounded by

$$\le \sup\{2eC(\varphi)\sqrt{\varepsilon}\|\Phi\|_{1,\varepsilon} - \|\Phi\|_{1,\varepsilon}^2\} = \text{const}(\sqrt{\varepsilon})^2 \longrightarrow 0. \qquad \square$$

Proof of Proposition 5.1. First note that

$$E^{\nu^{\varepsilon}} \left[B^{\varepsilon}(\varphi, Z)\Phi \right] = E^{\nu^{\varepsilon}} \left[\frac{Z(x)}{Z_{\rho}} \left(\{ \Psi^{\varepsilon} * \eta_2^{\varepsilon}(x) - \Psi^{\varepsilon}(x) \} - \frac{1}{12} \right) \Phi \right].$$

To compute this expectation, since $\{\Psi^{\varepsilon} * \eta_2^{\varepsilon}(x) - \Psi^{\varepsilon}(x)\}$ is a second-order Wiener functional, we need to pick up the second order and 0th order terms of the products of two Wiener functionals $\frac{Z(x)}{Z_{\rho}} \times \Phi$. We apply the diagram formula to compute the Wiener chaos expansion of products of two functions, cf. Major [179]. Note that we can expand under ν:

$$\frac{Z(x)}{Z_\rho} = e^{B(x) - \int_{\mathbb{R}} B(y)\rho(y)dy}$$

$$= e^{a(x)} \left\{ 1 + \sum_{n=1}^{\infty} \frac{1}{n!} \int_{\mathbb{R}^n} \phi_x^{\otimes n}(u_1, \ldots, u_n) dB(u_1) \cdots dB(u_n) \right\},$$

where

$$\phi_x(u) = 1_{(-\infty, x]}(u) - \int_u^{\infty} \rho(y)dy$$

and

$$a(x) = \frac{1}{2} \int_{\mathbb{R}} \phi_x(u)^2 du.$$

Note also that the kernel ϕ_x has a jump.

The constant $\frac{1}{24}$ is related to this jump and is actually given by $J/2$, where

$$J = P(R_1 + R_3 > 0, R_2 + R_3 > 0) - P(R_1 > 0, R_2 > 0),$$

and $\{R_i\}_{i=1}^3$ are i.i.d. random variables distributed under $\eta_2(x)dx$. Since η is symmetric,

$$P(R_1 + R_3 > 0, R_2 + R_3 > 0) = P(R_1 - R_3 > 0, R_2 - R_3 > 0)$$

$$= P(R_3 = \min R_i) = \frac{1}{3},$$

so that $J = \frac{1}{3} - \frac{1}{4} = \frac{1}{12}$. In this way, we obtain the term $\frac{1}{24}$. $\square$

Theorem 5.2 is shown under wrapping. The wrapping can be removed by showing a uniform estimate

$$\sup_{0<\varepsilon<1} E\left[\sup_{0 \le t \le T} h^\varepsilon(t, \rho)^2 \right] < \infty, \quad T > 0.$$

Namely, the height cannot move very fast. This is shown only on a torus, since we need to apply Poincaré's inequality. Thus, we obtain the SHE (5.13) on a torus $\mathbb{T}_M = \mathbb{R}/M\mathbb{Z}$ of arbitrary size M. However, letting $M \to \infty$, we finally obtain the SHE (5.13) on $\mathbb{R}$ in the limit under the stationarity assumption on the tilt processes.

The solutions $Z(t, x)$ of (5.13) and $Z_{CH}(t, x)$ of (5.5) are related under the simple transformation: $Z_{CH}(t, x) = e^{-\frac{t}{24}} Z(t, x)$. This leads to (5.14) by taking the log of the both sides.

Remark 5.2. 1. The constant $\frac{1}{24}$ is the speed of a growing interface and appears in many KPZ related papers. Bertini and Cancrini [25] discussed the moment Lyapunov exponents for the SHE (5.5), but with deterministic translation-invariant initial data, which are different from ours.

2. Hoshino [149] proved (5.14) in a non-stationary setting based on the paracontrolled calculus.

5.5 Invariant Measures of the Cole-Hopf Solution and SHE

As a byproduct of (5.14) combined with Theorem 5.1 and letting $\varepsilon \downarrow 0$, one can give a class of invariant measures for the stochastic heat equation (5.5) and for the Cole-Hopf solution of the KPZ equation. Let μ^c, $c \in \mathbb{R}$, be the distribution of the geometric Brownian motion $e^{B(x)+cx}$, $x \in \mathbb{R}$, with a drift c on $\mathscr{C}_+ = \{Z \in \mathscr{C}; Z > 0\}$, where $B(x)$ is the two-sided Brownian motion such that $\mu^c(B(0) \in dx) = dx$. Let ν^c be the distribution of $B(x)+cx$ on $\mathscr{C}$ such that $\nu^c(B(0) \in dx) = dx$. Note that these are not probability measures, in contrast to the measure ν on $\widetilde{\mathscr{C}}$ introduced in the preceding section, which is a probability measure.

Theorem 5.3 ([112]). *The measures $\{\mu^c\}_{c\in\mathbb{R}}$ are invariant under the SHE (5.5), i.e., if $Z(0) \overset{\text{law}}{=} \mu^c$, then $Z(t) \overset{\text{law}}{=} \mu^c$ for all $t \geq 0$ and $c \in \mathbb{R}$.*

Corollary 5.1. *The measures $\{\nu^c\}_{c\in\mathbb{R}}$ are invariant under the Cole-Hopf solution of the KPZ equation.*

The constant c represents the average tilt of the interface. We have different invariant measures for different average tilts. Reversibility does not hold, but a kind of Yaglom reversibility holds. The SHE (5.5) has the *scale invariance*: If $Z(t, x)$ is a solution of (5.5), then

$$Z^c(t, x) := e^{cx+\frac{1}{2}c^2t}Z(t, x + ct)$$

is also a solution (with a new space-time white noise). Therefore, once the invariance of μ^0 is shown, μ^c is also invariant, for every $c \in \mathbb{R}$.

As we mentioned above, (5.14) combined with Theorem 5.1 at the approximating level shows the invariance of μ^0 for tilt processes. To extend this to the height processes $Z(t)$, we introduce the transformation $h^\varepsilon(x, Z) := \log(Z * \eta^\varepsilon(x))$. Then, Theorem 5.3 is shown by noting that the evolution of $h^\varepsilon(x, Z(t))$ is governed only by the tilt variables and the initial data $h^\varepsilon(x, Z(0))$.

One expects μ^c, $c \in \mathbb{R}$, to be all the extremal invariant measures (up to constant multipliers), but this remains open, cf. Funaki and Spohn [116] for such a result for the $\nabla\varphi$-interface model.

5.6   Multi-component Coupled KPZ Equation

Motivated by the problems in fluctuating hydrodynamics mentioned below, Ferrari, Sasamoto, and Spohn [79] studied an $\mathbb{R}^d$-valued coupled KPZ equation for $h(t, x) = (h^\alpha(t, x))_{\alpha=1}^d$ on $\mathbb{T}$ (or $\mathbb{R}$):

$$\partial_t h^\alpha = \frac{1}{2}\partial_x^2 h^\alpha + \frac{1}{2}\Gamma_{\beta\gamma}^\alpha \partial_x h^\beta \partial_x h^\gamma + \sigma_\beta^\alpha \dot{W}^\beta, \ x \in \mathbb{T}, \tag{5.18}$$

where we use Einstein's convention, $\dot{W}(t, x) = (\dot{W}^\alpha(t, x))_{\alpha=1}^d$ is an $\mathbb{R}^d$-valued space-time Gaussian white noise with the covariance

$$E[\dot{W}^\alpha(t, x)\dot{W}^\beta(s, y)] = \delta^{\alpha\beta}\delta(x - y)\delta(t - s),$$

and $\delta^{\alpha\beta}$ denotes Kronecker's δ. A similar SPDE appears in the study of random motion of loops on a manifold, cf. Funaki [94]. The constants $(\sigma_\beta^\alpha)_{1\le\alpha,\beta\le d}$ and $(\Gamma_{\beta\gamma}^\alpha)_{1\le\alpha,\beta,\gamma\le d}$ are given and, from the form of the equation (5.18), $\Gamma_{\beta\gamma}^\alpha$ ought to satisfy

$$\Gamma_{\beta\gamma}^\alpha = \Gamma_{\gamma\beta}^\alpha, \tag{5.19}$$

for all α, β, γ.

Two approximating equations: Symmetric convolution kernels $\eta \in C_0^\infty(\mathbb{R})$ and $\eta^\varepsilon, \eta_2^\varepsilon$ are given similarly as before. A simple $\mathbb{R}^d$-valued KPZ approximating equation with a smeared noise is given by

$$\partial_t h^\alpha = \frac{1}{2}\partial_x^2 h^\alpha + \frac{1}{2}\Gamma_{\beta\gamma}^\alpha (\partial_x h^\beta \partial_x h^\gamma - c^\varepsilon A^{\beta\gamma} - B^{\varepsilon,\beta\gamma}) + \sigma_\beta^\alpha \dot{W}^\beta * \eta^\varepsilon, \tag{5.20}$$

where $A^{\beta\gamma} = \sum_{\delta=1}^d \sigma_\delta^\beta \sigma_\delta^\gamma$ and the renormalization factor $c^\varepsilon = \frac{1}{\varepsilon}\|\eta\|_{L^2(\mathbb{R})}^2$ is the same as before. We need another scaling factor $B^{\varepsilon,\beta\gamma}$ which behaves as $O(-\log\varepsilon)$ in general.

As another approximation suitable for studying invariant measures, we consider the second $\mathbb{R}^d$-valued KPZ approximating equation for $h = h^\varepsilon(t, x) \equiv (h^{\varepsilon,\alpha}(t, x))_{\alpha=1}^d$:

$$\partial_t h^\alpha = \frac{1}{2}\partial_x^2 h^\alpha + \frac{1}{2}\Gamma_{\beta\gamma}^\alpha (\partial_x h^\beta \partial_x h^\gamma - c^\varepsilon A^{\beta\gamma} - \tilde{B}^{\varepsilon,\beta\gamma}) * \eta_2^\varepsilon + \sigma_\beta^\alpha \dot{W}^\beta * \eta^\varepsilon, \tag{5.21}$$

where $\tilde{B}^{\varepsilon,\beta\gamma}$ is another scaling factor. We now assume the matrix $\sigma = (\sigma_\beta^\alpha)_{1\le\alpha,\beta\le d}$ is invertible and denote its inverse matrix by $\tau = (\tau_\beta^\alpha)_{1\le\alpha,\beta\le d}$, i.e., $\sigma_\beta^\alpha \tau_\gamma^\beta = \delta_\gamma^\alpha$. We define $\hat{\Gamma}_{\beta\gamma}^\alpha$ as

$$\hat{\Gamma}^{\alpha}_{\beta\gamma} := \tau^{\alpha}_{\alpha'}\Gamma^{\alpha'}_{\beta'\gamma'}\sigma^{\beta'}_{\beta}\sigma^{\gamma'}_{\gamma}. \tag{5.22}$$

Note that the relation $\hat{\Gamma}^{\alpha}_{\beta\gamma} = \hat{\Gamma}^{\alpha}_{\gamma\beta}$ holds also for $\hat{\Gamma}$. Then, for the solution of the second $\mathbb{R}^d$-valued KPZ approximating equation (5.21), under the additional condition

$$\hat{\Gamma}^{\alpha}_{\beta\gamma} = \hat{\Gamma}^{\gamma}_{\alpha\beta}, \tag{5.23}$$

one can show the infinitesimal invariance of the distribution of $(\sigma B) * \eta^{\varepsilon}(x)$, where $B = (B^{\alpha})_{\alpha}$ is the $\mathbb{R}^d$-valued two-sided Brownian motion and $\sigma B = (\sigma^{\alpha}_{\beta}B^{\beta})^d_{\alpha=1}$:

Theorem 5.4. *The distribution ν^{ε} of $\partial_x((\sigma B) * \eta^{\varepsilon}(x))$ on $\mathscr{C} = C(\mathbb{R}, \mathbb{R}^d)$ is infinitesimally invariant for the tilt process $\partial_x h$ determined by the SPDE (5.21).*

Note that ν^{ε} does not depend on the choice of $B(0)$. This theorem was shown by Funaki [107] on $\mathbb{R}$ when $\sigma = I$ (the identity matrix), one can easily extend it to our setting. When $d = 1$ and $\Gamma^{\alpha}_{\beta\gamma} = \sigma^{\alpha}_{\beta} = 1$, the approximating equations (5.20) and (5.21) are the same as (5.8) and (5.10), respectively, since we can take $B^{\varepsilon,\beta\gamma} = \tilde{B}^{\varepsilon,\beta\gamma} = 0$ in the scalar valued case.

Passing to the limit $\varepsilon \downarrow 0$: The next goal is to pass to the limit $\varepsilon \downarrow 0$. As we saw in Sect. 5.4, the solution of the first KPZ approximating equation (5.8) converges as $\varepsilon \downarrow 0$ to the Cole-Hopf solution $h_{\mathrm{CH}}(t, x)$ of the KPZ equation, while the solution of the second KPZ approximating equation (5.10) converges to $h_{\mathrm{CH}}(t, x) + \frac{1}{24}t$. The method used in Sect. 5.4 relies heavily on the Cole-Hopf transformation, which in general works only for the scalar-valued equations and not for the multi-component coupled equations. Instead, to investigate the limits of the solutions of the two $\mathbb{R}^d$-valued KPZ approximating equations (5.20) and (5.21) as $\varepsilon \downarrow 0$, we apply the *paracontrolled calculus* due to Gubinelli and others [130, 131]. In particular, we study the difference between the limits of these two equations.

For $\kappa \in \mathbb{R}$, $\mathscr{C}^{\kappa}(= \mathscr{B}^{\kappa}_{\infty,\infty}(\mathbb{T}))$ denotes the $\mathbb{R}^d$-valued Besov space on $\mathbb{T}$.

Theorem 5.5 (Funaki and Hoshino [108]). 1. *Assume $h(0) \in \bigcup_{\delta>0} \mathscr{C}^{\delta}$. Then the solution h^{ε} of (5.20) with a proper choice of $B^{\varepsilon,\beta\gamma}$ converges in probability as $\varepsilon \downarrow 0$ to some h in $C([0, T], \mathscr{C}^{\frac{1}{2}-\delta})$ for every $\delta > 0$ and $t \in [0, T]$ with some $T > 0$.*
2. *A similar result holds for the solution $\tilde{h}^{\varepsilon}$ of (5.21) and it converges to some $\tilde{h}$.*
3. *We assume the condition (5.23) in addition to (5.19). Then, both $B^{\varepsilon,\beta\gamma}$ and $\tilde{B}^{\varepsilon,\beta\gamma}$ behaves as $O(1)$, so that the solutions of (5.20) with $B^{\varepsilon,\beta\gamma} = 0$ and (5.21) with $\tilde{B}^{\varepsilon,\beta\gamma} = 0$ converge. In the limit, we have*

$$\tilde{h}^{\alpha}(t, x) = h^{\alpha}(t, x) + c^{\alpha}t, \quad 1 \leq \alpha \leq d,$$

where

$$c^\alpha = \frac{1}{24} \sum_{\gamma,\gamma'} \sigma_\beta^\alpha \hat{\Gamma}_{\alpha'\alpha''}^\beta \hat{\Gamma}_{\gamma\gamma'}^{\alpha'} \hat{\Gamma}_{\gamma\gamma'}^{\alpha''}.$$

This theorem is an extension of (5.14) from the scalar-valued case to the multi-component and non-stationary setting.

We can also see that, if the condition (5.23) ensuring the infinitesimal invariance is violated, both $B^{\varepsilon,\beta\gamma}$ and $\tilde{B}^{\varepsilon,\beta\gamma}$ diverge.

Fluctuating hydrodynamics: From microscopic systems with random evolutions, via the space-time scaling passing through the so-called local equilibrium (or a local average under local ergodicity), one can derive certain nonlinear PDEs. This procedure is called the hydrodynamic limit and one example was given in Sect. 2.2.2. If the system has d (local) conserved quantities (e.g., the three-dimensional compressible Euler equation in fluid dynamics has five conserved quantities, mass, momenta and energy), then in the limit we have a system of d coupled nonlinear PDEs. The noises (random fluctuations) in the microscopic systems are averaged out and disappear in macroscopic limit equations.

However, if we consider the linearization of this system around a global equilibrium, the noise terms survive under a proper scaling and we obtain linear SPDEs in the limit, as we discussed in Sect. 2.2.3. At least heuristically, if we expand the equation to the second order, we can expect to obtain the coupled KPZ equations in the limit.

A physically interesting object is the covariance structure of the solution of the coupled KPZ equation. One can expect the behavior like

$$\langle h(t,x); h(0,0) \rangle_{\text{equil}} \asymp (ct)^{-\kappa} f\big((ct)^{-\kappa} x\big),$$

under equilibrium, with some function f and $\kappa > 0$.

Depending on κ, we call

1. Diffusive scaling iff $\kappa = \frac{1}{2}$, $f = e^{-cx^2}$,
2. KPZ scaling iff $\kappa = \frac{2}{3}$, $f = f_{\text{KPZ}}$,
3. α-Lévy scaling iff $\kappa = \frac{1}{\alpha}$ with $0 < \alpha < 2$. This is sometimes called an anomalous behavior, for example, with $\alpha = \frac{3}{5}, \kappa = \frac{5}{3}$.

For the coupled KPZ equation, for example for $d = 2$, if $\Gamma_{11}^1 \neq 0$ and $\Gamma_{22}^2 \neq 0$, both h^1 and h^2 have KPZ scalings. On the other hand, if $\Gamma_{\alpha\beta}^\gamma = 0$ for all α, β, γ, then the system is linear and decoupled, and the scaling should be diffusive. A mixed case such as $\Gamma_{11}^1 \neq 0$ and $\Gamma_{22}^2 = \Gamma_{11}^2 = 0$ is interesting. It is expected that two different modes coexist. See Spohn et al. [206, 207] for details.

References

1. Alfaro, M., Antonopoulou, D., Karali, G., Matano, H.: Generation and propagation of fine transition layers for the stochastic Allen-Cahn equation (2016, preprint)
2. Alfaro, M., Garcke, H., Hilhorst, D., Matano, H., Schätzle, R.: Motion by anisotropic mean curvature as sharp interface limit of an inhomogeneous and anisotropic Allen-Cahn equation. Proc. R. Soc. Edinb. Sect. A **140**, 673–706 (2010)
3. Alfaro, M., Hilhorst, D., Matano, H.: The singular limit of the Allen-Cahn equation and the FitzHugh-Nagumo system. J. Differ. Equs. **245**, 505–565 (2008)
4. Alfaro, M., Matano, H.: On the validity of formal asymptotic expansions in Allen-Cahn equation and FitzHugh-Nagumo system with generic initial data. Discret. Contin. Dyn. Syst. Ser. B **17**, 1639–1649 (2012)
5. Alikakos, N.D., Bates, P.W., Chen, X.: Convergence of the Cahn-Hilliard equation to the Hele-Shaw model. Arch. Ration. Mech. Anal. **128**, 165–205 (1994)
6. Allen, S.M., Cahn, J.W.: A macroscopic theory for antiphase boundary motion and its application to antiphase domain coarsening. Acta. Metall. **27**, 1085–1095 (1979)
7. Almgren, F., Taylor, J.E., Wang, L.: Curvature-driven flows: a variational approach. SIAM J. Control Optim. **31**, 387–438 (1993)
8. Alt, H.W., Caffarelli, L.A.:. Existence and regularity for a minimum problem with free boundary. J. Reine Angew. Math. **325**, 105–144 (1981)
9. Alt, H.W., Caffarelli, L.A., Friedman, A.: Variational problems with two phases and their free boundaries. Trans. Am. Math. Soc. **282**, 431–461 (1984)
10. Amir, G., Corwin, I., Quastel, J.: Probability distribution of the free energy of the continuum directed random polymer in $1 + 1$ dimensions. Commun. Pure Appl. Math. **64**, 466–537 (2011)
11. Andjel, E.D.: Invariant measures for the zero range processes. Ann. Probab. **10**, 525–547 (1982)
12. Angenent, S.B.: Some recent results on mean curvature flow. In: Recent Advances in Partial Differential Equations. RAM Recherches en mathématiques appliquées, vol. 30, pp. 1–18. Masson, Paris (1994)
13. Antonopoulou, D., Blömker, D., Karali, G.: Front motion in the one-dimensional stochastic Cahn-Hilliard equation. SIAM J. Math. Anal. **44**, 3242–3280 (2012)
14. Antonopoulou, D., Karali, G., Kossioris, G.: Asymptotics for a generalized Cahn-Hilliard equation with forcing terms. Discret. Contin. Dyn. Syst. Ser. A **30**, 1037–1054 (2011)
15. Aronson, D.G., Weinberger, H.F.: Multidimensional nonlinear diffusion arising in population genetics. Adv. Math. **30**, 33–76 (1978)

16. Baccelli, F., Karpelevich, F.I., Kelbert, M.Ya., Puhalskii, A.A., Rybko, A.N., Suhov, Yu.M.: A mean-field limit for a class of queueing networks. J. Stat. Phys. **66**, 803–825 (1992)

17. Balázs, M., Quastel, J., Seppäläinen, T.: Fluctuation exponent of the KPZ/stochastic Burgers equation. J. Am. Math. Soc. **24**, 683–708 (2011)

18. Barles, G., Souganidis, P.E.: A new approach to front propagation problems: theory and applications. Arch. Ration. Mech. Anal. **141**, 237–296 (1998)

19. Bellettini, G.: Lecture Notes on Mean Curvature Flow, Barriers and Singular Perturbations. Lecture Notes, vol. 12. Scuola Normale, Superiore di Pisa (2013)

20. Beltoft, D., Boutillier, C., Enriquez, N.: Random young diagrams in a rectangular box. Mosc. Math. J. **12**, 719–745 (2012)

21. Bertini, L., Brassesco, S., Buttà, P., Presutti, E.: Front fluctuations in one dimensional stochastic phase field equations. Ann. Henri Poincaré **3**, 29–86 (2002)

22. Bertini, L., Brassesco, S., Buttà, P.: Soft and hard wall in a stochastic reaction diffusion equation. Arch. Ration. Mech. Anal. **190**, 307–345 (2008)

23. Bertini, L., Brassesco, S., Buttà, P.: Front fluctuations for the stochastic Cahn-Hilliard equation. Braz. J. Probab. Stat. **29**, 336–371 (2015)

24. Bertini, L., Buttà, P., Pisante, A.: Stochastic Allen-Cahn approximation of the mean curvature flow: large deviations upper bound. arXiv:1604.02064

25. Bertini, L., Cancrini, N.: The stochastic heat equation: Feynman-Kac formula and intermittence. J. Stat. Phys. **78**, 1377–1401 (1995)

26. Bertini, L., De Sole, A., Gabrielli, D., Jona-Lasinio, G., Landim, C.: Large deviation approach to non equilibrium processes in stochastic lattice gases. Bull. Braz. Math. Soc. **37**, 611–643 (2006)

27. Bertini, L., Giacomin, G.: Stochastic Burgers and KPZ equations from particle systems. Commun. Math. Phys. **183**, 571–607 (1997)

28. Bertini, L., Landim, C., Mourragui, M.: Dynamical large deviations for the boundary driven weakly asymmetric exclusion process. Ann. Probab. **37**, 2357–2403 (2009)

29. Bodineau, T.: The Wulff construction in three and more dimensions. Commun. Math. Phys. **207**, 197–229 (1999)

30. Bodineau, T., Ioffe, D., Velenik, Y.: Rigorous probabilistic analysis of equilibrium crystal shapes, probabilistic techniques in equilibrium and nonequilibrium statistical physics. J. Math. Phys. **41**, 1033–1098 (2000)

31. Bolthausen, E., Chiyonobu, T., Funaki, T.: Scaling limits for weakly pinned Gaussian random fields under the presence of two possible candidates. J. Math. Soc. Jpn. **67**, 1359–1412 (2015) (special issue for Kiyosi Itô)

32. Bolthausen, E., Funaki, T., Otobe, T.: Concentration under scaling limits for weakly pinned Gaussian random walks. Probab. Theory Relat. Fields **143**, 441–480 (2009)

33. Bolthausen, E., Ioffe, D.: Harmonic crystal on the wall: a microscopic approach. Commun. Math. Phys. **187**, 523–566 (1997)

34. Borodin, A., Corwin, I., Ferrari, P., Veto, B.: Height fluctuations for the stationary KPZ equation. Math. Phys. Anal. Geom. **18**(Art. 20), 95 (2015)

35. Bounebache, S.K.: A random string with reflection in a convex domain. Stoch. Anal. Appl. **29**, 523–549 (2011)

36. Bounebache, S.K., Zambotti, L.: A skew stochastic heat equation. J. Theor. Probab. **27**, 168–201 (2014)

37. Boutillier, C.: Pattern densities in non-frozen planar dimer models. Commun. Math. Phys. **271**, 55–91 (2007)

38. Brakke, K.A.: The Motion of a Surface by its Mean Curvature. Mathematical Notes, vol. 20. Princeton University Press, Princeton (1978)

39. Brassesco, S., Buttà, P.: Interface fluctuations for the $D = 1$ stochastic Ginzburg-Landau equation with nonsymmetric reaction term. J. Stat. Phys. **93**, 1111–1142 (1998)

40. Brassesco, S., De Masi, A., Presutti, E.: Brownian fluctuations of the instanton in the $d = 1$ Ginzburg-Landau equation with noise. Ann. Inst. H. Poincaré Probab. Statist. **31**, 81–118 (1995)

41. Cahn, J.W., Elliott, C.M., Novick-Cohen, A.: The Cahn-Hilliard equation with a concentration dependent mobility: motion by minus the Laplacian of the mean curvature Eur. J. Appl. Math. **7**, 287–301 (1996)

42. Caputo, P., Martinelli, F., Simenhaus, F., Toninelli, F.: "Zero" temperature stochastic 3D Ising model and dimer covering fluctuations: a first step towards interface mean curvature motion. Commun. Pure Appl. Math. **64**, 778–831 (2011)

43. Caputo, P., Martinelli, F., Toninelli, F.: Mixing times of monotone surfaces and SOS interfaces: a mean curvature approach. Commun. Math. Phys. **311**, 157–189 (2012)

44. Carr, J., Pego, R.L.: Metastable patterns in solutions of $u_t = \epsilon^2 u_{xx} - f(u)$. Commun. Pure Appl. Math. **42**, 523–576 (1989)

45. Cerf, R., Kenyon, R.: The low-temperature expansion of the Wulff crystal in the 3D Ising model. Commun. Math. Phys. **222**, 147–179 (2001)

46. Cerf, R., Pisztora, A.: On the Wulff crystal in the Ising model. Ann. Probab. **28**, 947–1017 (2000)

47. Chandra, A., Weber, H.: Stochastic PDEs, regularity structures, and interacting particle systems. arXiv:1508.03616

48. Chang, C.C., Yau, H.-T.: Fluctuations of one dimensional Ginzburg-Landau models in nonequilibrium. Commun. Math. Phys. **145**, 209–239 (1992)

49. Chen, X.: Generation and propagation of interfaces for reaction-diffusion equations. J. Differ. Equs. **96**, 116–141 (1992)

50. Chen, X.: Spectrum for the Allen-Cahn, Cahn-Hilliard, and phase-field equations for generic interfaces. Commun. Part. Differ. Equs. **19**, 1371–1395 (1994)

51. Chen, X.: Generation, propagation, and annihilation of metastable patterns. J. Differ. Equs. **206**, 399–437 (2004)

52. Chen, X., Hilhorst, D., Logak, E.: Mass conserving Allen-Cahn equation and volume preserving mean curvature flow. Interfaces Free Bound. **12**, 527–549 (2010)

53. Chen, Y.G., Giga, Y., Goto, S.: Uniqueness and existence of viscosity solutions of generalized mean curvature flow equations. J. Differ. Geom. **33**, 749–786 (1991)

54. Corwin, I., Tsai, L.-C.: KPZ equation limit of higher-spin exclusion processes. Ann. Probab. (2016, published online)

55. Corwin, I., Shen, H., Tsai, L.-C.: ASEP(q, j) converges to the KPZ equation. arXiv:1602.01908

56. Da Prato, G. , Debussche, A., Tubaro, L.: A modified Kardar-Parisi-Zhang model. Electron. Commun. Probab. **12**, 442–453 (2007)

57. Da Prato, G., Zabczyk, J.: Stochastic Equations in Infinite Dimensions. Encyclopedia of Mathematics and its Applications. Cambridge University Press, Cambridge/New York (1992)

58. Da Prato, G., Zabczyk, J.: Ergodicity for Infinite Dimensional Systems. London Mathematical Society Lecture Notes, vol. 229. Cambridge University Press, Cambridge/New York (1996)

59. Debussche, A., Zambotti, L.: Conservative stochastic Cahn-Hilliard equation with reflection. Ann. Probab. **35**, 1706–1739 (2007)

60. De Masi, A., Presutti, E., Scacciatelli, E.: The weakly asymmetric simple exclusion process. Ann. Inst. H. Poincaré Probab. Statist. **25**, 1–38 (1989)

61. Dembo, A., Vershik, A., Zeitouni, O.: Large deviations for integer partitions. Markov Process. Related Fields **6**, 147–179 (2000)

62. Dembo, A., Zeitouni, O.: Large Deviations Techniques and Applications, 2nd edn. Applications of Mathematics, vol. 38. Springer, New York (1998)

63. Deuschel, J.-D., Giacomin, G., Ioffe, D.: Large deviations and concentration properties for $\nabla \varphi$ interface models. Probab. Theory Relat. Fields **117**, 49–111 (2000)

64. Dirr, N., Luckhaus, S., Novaga, M.: A stochastic selection principle in case of fattening for curvature flow. Calc. Var. Part. Differ. Equs. **13**, 405–425 (2001)

65. Dobrushin, R.L., Kotecký, R., Shlosman, S.: Wulff Construction: A Global Shape from Local Interaction. AMS Translation Series, vol. 104. American Mathematical Society, Providence (1992)

66. Duits, M.: Gaussian free field in an interlacing particle system with two jump rates. Commun. Pure Appl. Math. **66**, 600–643 (2013)
67. Ecker, K., Huisken, G.: Mean curvature evolution of entire graphs. Ann. Math. **130**, 453–471 (1989)
68. Elliott, C.M., Garcke, H.: Existence results for diffusive surface motion laws. Adv. Math. Sci. Appl. **7**, 467–490 (1997)
69. Elworthy, K.D., Truman, A., Zhao, H.Z., Gaines, J.G.: Approximate travelling waves for generalized KPP equations and classical mechanics. Proc. R. Soc. Lond. A **446**, 529–554 (1994)
70. Es-Sarhir, A., von Renesse, M., Stannat, W.: Estimates for the ergodic measure and polynomial stability of plane stochastic curve shortening flow. Nonlinear Differ. Equ. Appl. **19**, 663–675 (2012)
71. Es-Sarhir, A., von Renesse, M.: Ergodicity of stochastic curve shortening flow in the plane. SIAM J. Math. Anal. **44**, 224–244 (2012)
72. Escher, J., Simonett, G.: The volume preserving mean curvature flow near spheres. Proc. Am. Math. Soc. **126**, 2789–2796 (1998)
73. Evans, L.C., Soner, H.M., Souganidis, P.E.: Phase transitions and generalized motion by mean curvature. Commun. Pure Appl. Math. **45**, 1097–1123 (1992)
74. Evans, L.C., Spruck, J.: Motion of level sets by mean curvature I. J. Differ. Geom. **33**, 635–681 (1991)
75. Evans, L.C., Spruck, J.: Motion of level sets by mean curvature. II. Trans. Am. Math. Soc. **330**, 321–332 (1992)
76. Eyink, G., Lebowitz, J.L., Spohn, H.: Lattice gas models in contact with stochastic reservoirs: local equilibrium and relaxation to the steady state. Commun. Math. Phys. **140**, 119–131 (1991)
77. Fatkullin, I., Kovačič, G., Vanden-Eijnden, E.: Reduced dynamics of stochastically perturbed gradient flows. Commun. Math. Sci. **8**, 439–461 (2010)
78. Feng, X., Prohl, A.: Numerical analysis of the Allen-Cahn equation and approximation for mean curvature flows. Numer. Math. **94**, 33–65 (2003)
79. Ferrari, P.L., Sasamoto, T., Spohn, H.: Coupled Kardar-Parisi-Zhang equations in one dimension. J. Stat. Phys. **153**, 377–399 (2013)
80. Ferrari, P.L., Spohn, H.: Step fluctuations for a faceted crystal. J. Stat. Phys. **113**, 1–46 (2003)
81. Fife, P.C., Hsiao, L.: The generation and propagation of internal layers. Nonlinear Anal. **12**, 19–41 (1988)
82. Fife, P.C., McLeod, J.B.: The approach of solutions of nonlinear diffusion equations to travelling front solutions. Arch. Rat. Mech. Anal. **65**, 335–361 (1977)
83. Freidlin, M.: Functional Integration and Partial Differential Equations. Princeton University Press, Princeton (1985)
84. Freidlin, M.: Semi-linear PDE's and limit theorems for large deviations. In: Hennequin (ed.) Lectures on Probability Theory and Statistics: Ecole d'Eté de Probabilités de Saint-Flour XX – 1990. Lecture Notes in Mathematics, vol. 1527, pp. 2–109. Springer (1992)
85. Freiman, G., VERSHIK, A., YAKUBOVICH, Y.: A local limit theorem for random strict partitions. Theory Probab. Appl. **44**, 453–468 (2000)
86. Friedrichs, K.: Über ein Minimumproblem für Potentialströmungen mit freiem Rande. Math. Ann. **109**, 60–82 (1934)
87. Fritz, J.: On the diffusive nature of entropy flow in infinite systems: remarks to a paper by Guo-Papanicolau-Varadhan. Commun. Math. Phys. **133**, 331–352 (1990)
88. Friz, P.K., Hairer, M.: A Course on Rough Paths. With an Introduction to Regularity Structures. Universitext. Springer, Cham (2014)
89. Funaki, T.: Random motion of strings and related stochastic evolution equations. Nagoya Math. J. **89**, 129–193 (1983)
90. Funaki, T.: Derivation of the hydrodynamical equation for one-dimensional Ginzburg-Landau model. Probab. Theory Relat. Fields **82**, 39–93 (1989)
91. Funaki, T.: The hydrodynamic limit for a system with interactions prescribed by Ginzburg-Landau type random Hamiltonian. Probab. Theory Relat. Fields **90**, 519–562 (1991)

92. Funaki, T.: The reversible measures of multi-dimensional Ginzburg-Landau type continuum model. Osaka J. Math. **28**, 463–494 (1991)
93. Funaki, T.: Regularity properties for stochastic partial differential equations of parabolic type. Osaka J. Math. **28**, 495–516 (1991)
94. Funaki, T.: A stochastic partial differential equation with values in a manifold. J. Funct. Anal. **109**, 257–288 (1992)
95. Funaki, T.: Low temperature limit and separation of phases for Ginzburg-Landau stochastic equation. In: Kunita and Kuo (eds.) Stochastic Analysis on Infinite Dimensional Spaces, Proceedings of the U.S.-Japan Bilateral Seminar at Baton Rouge. Pitman Research Notes in Mathematical Series, vol. 310, pp. 88–98. Longman, Essex (1994)
96. Funaki, T.: The scaling limit for a stochastic PDE and the separation of phases. Probab. Theory Relat. Fields **102**, 221–288 (1995)
97. Funaki, T.: Singular limit for reaction-diffusion equation with self-similar Gaussian noise. In: Elworthy, Kusuoka, Shigekawa (eds.) Proceedings of Taniguchi Symposium, New Trends in Stochastic Analysis, pp. 132–152. World Scientific (1997)
98. Funaki, T.: Singular limit for stochastic reaction-diffusion equation and generation of random interfaces. Acta Math. Sinica Engl. Ser. **15**, 407–438 (1999)
99. Funaki, T.: Hydrodynamic limit for $\nabla\phi$ interface model on a wall. Probab. Theory Relat. Fields **126**, 155–183 (2003)
100. Funaki, T.: Stochastic models for phase separation and evolution equations of interfaces. Sugaku Expositions **16**, 97–116 (2003)
101. Funaki, T.: Zero temperature limit for interacting Brownian particles, I. Motion of a single body. Ann. Probab. **32**, 1201–1227 (2004)
102. Funaki, T.: Zero temperature limit for interacting Brownian particles, II. Coagulation in one dimension. Ann. Probab. **32**, 1228–1246 (2004)
103. Funaki, T.: Stochastic interface models. In: Picard, J. (ed.) Lectures on Probability Theory and Statistics, Ecole d'Eté de Probabilités de Saint-Flour XXXIII – 2003. Lecture Notes in Mathematics, vol. 1869, pp. 103–274. Springer, Berlin (2005)
104. Funaki, T.: Stochastic Differential Equations (in Japanese) Iwanami, 1997, 203p (2005). xviii+187 pages
105. Funaki, T.: Stochastic analysis on large scale interacting systems. In: Selected Papers on Probability and Statistics. Am. Math. Soc. Trans. Ser. 2 **227**, 49–73 (2009)
106. Funaki, T.: Equivalence of ensembles under inhomogeneous conditioning and its applications to random Young diagrams. J. Stat. Phys. **154**, 588–609 (2014) (special issue for Herbert Spohn)
107. Funaki, T.: Infinitesimal invariance for the coupled KPZ equations, Memoriam Marc Yor – Séminaire de Probabilités XLVII. Lecture Notes in Mathematics, vol. 2137, pp. 37–47. Springer (2015)
108. Funaki, T., Hoshino, M.: A coupled KPZ equation, its two types of approximations and existence of global solutions. arXiv:1611.00498
109. Funaki, T., Nishikawa, T.: Large deviations for the Ginzburg-Landau $\nabla\phi$ interface model. Probab. Theory Related Fields **120**, 535–568 (2001)
110. Funaki, T., Olla, S.: Fluctuations for $\nabla\phi$ interface model on a wall. Stoch. Proc. Appl. **94**, 1–27 (2001)
111. Funaki, T., Otobe, T.: Scaling limits for weakly pinned random walks with two large deviation minimizers. J. Math. Soc. Jpn. **62**, 1005–1041 (2010)
112. Funaki, T., Quastel, J.: KPZ equation, its renormalization and invariant measures. Stoch. Partial Differ. Equ. Anal. Comput. **3**, 159–220 (2015)
113. Funaki, T., Sakagawa, H.: Large deviations for $\nabla\varphi$ interface model and derivation of free boundary problems. In: Funaki, T., Osada, H. (eds.) Stochastic Analysis on Large Scale Interacting Systems. Advanced Studies in Pure Mathematics, vol. 39, pp. 173–211. Mathematical Society of Japan, Tokyo (2004)
114. Funaki, T., Sasada, M.: Hydrodynamic limit for an evolutional model of two-dimensional Young diagrams. Commun. Math. Phys. **299**, 335–363 (2010)

115. Funaki, T., Sasada, M., Sauer, M., Xie, B.: Fluctuations in an evolutional model of two-dimensional Young diagrams. Stoch. Proc. Appl. **123**, 1229–1275 (2013)
116. Funaki, T., Spohn, H.: Motion by mean curvature from the Ginzburg-Landau $\nabla\phi$ interface model. Commun. Math. Phys. **185**, 1–36 (1997)
117. Funaki, T., Xie, B.: A stochastic heat equation with the distributions of Lévy processes as its invariant measures. Stoch. Proc. Appl. **119**, 307–326 (2009)
118. Funaki, T., Yokoyama, S.: Sharp interface limit for stochastically perturbed mass conserving Allen-Cahn equation. arXiv:1610.01263
119. Gage, M., Hamilton, R.S.: The heat equation shrinking convex plane curves. J. Differ. Geom. **23**, 69–96 (1986)
120. Gärtner, J.: Bistable reaction-diffusion equations and excitable media. Math. Nachr. **112**, 125–152 (1983)
121. Gärtner, J.: Convergence towards Burger's equation and propagation of chaos for weakly asymmetric exclusion processes. Stoch. Process. Appl. **27**, 233–260 (1988)
122. Giga, Y.: Surface Evolution Equations. A Level Set Approach. Monographs in Mathematics, vol. 99. Birkhäuser, Basel/Boston (2006)
123. Giga, Y., Mizoguchi, N.: Existence of periodic solutions for equations of evolving curves. SIAM J. Math. Anal. **27**, 5–39 (1996)
124. Glimm, J., Jaffe, A.: Quantum Physics. A Functional Integral Point of View, 2nd edn. Springer, New York (1987)
125. Glimm, J., Jaffe, A., Spencer, T.: Phase transitions for φ_2^4 quantum fields. Commun. Math. Phys. **45**, 203–216 (1975)
126. Glimm, J., Jaffe, A., Spencer, T.: Phase transitions in $P(\phi)_2$ quantum fields. Bull. Am. Math. Soc. **82**, 713–715 (1976)
127. Gonçalves, P., Jara, M.: Nonlinear fluctuations of weakly asymmetric interacting particle systems. Arch. Ration. Mech. Anal. **212**, 597–644 (2014)
128. Gonçalves, P., Jara, M., Sethuraman, S.: A stochastic Burgers equation from a class of microscopic interactions. Ann. Probab. **43**, 286–338 (2015)
129. Grayson, M.A.: The heat equation shrinks embedded plane curves to round points. J. Differ. Geom. **26**, 285–314 (1987)
130. Gubinelli, M., Imkeller, P., Perkowski, N.: Paracontrolled distributions and singular PDEs. Forum Math. Pi **3**(e6), 75 (2015)
131. Gubinelli, M., Perkowski, N.: KPZ reloaded. arXiv:1508.03877
132. Gubinelli, M., Perkowski, N.: Energy solutions of KPZ are unique. arXiv:1508.07764
133. Gubinelli, M., Perkowski, N.: The Hairer-Quastel universality result in equilibrium. RIMS Kôkyûroku Bessatsu **B59**, 101–115 (2016)
134. Guo, M.Z., Papanicolaou, G.C., Varadhan, S.R.S.: Nonlinear diffusion limit for a system with nearest neighbor interactions. Commun. Math. Phys. **118**, 31–59 (1988)
135. Gurtin, M.E.: Thermomechanics of Evolving Phase Boundaries in the Plane. Clarendon Press, Oxford (1993)
136. Hadeler, K.P., Rothe, F.: Travelling fronts in nonlinear diffusion equations. J. Math. Biol. **2**, 251–263 (1975)
137. Hairer, M.: Ergodic Theory for Stochastic PDEs (2008). Available online at http://www.hairer.org/notes/Imperial.pdf
138. Hairer, M.: Solving the KPZ equation. Ann. Math. **178**, 559–664 (2013)
139. Hairer, M.: A theory of regularity structures. Invent. Math. **198**, 269–504 (2014)
140. Hairer, M., Matetski, K.: Discretisations of rough stochastic PDEs. arXiv:1511.06937
141. Hairer, M., Mattingly, J.: Ergodicity of the 2D Navier-Stokes equations with degenerate stochastic forcing. Ann. Math. **164**, 993–1032 (2006)
142. Hairer, M., Pardoux, E.: A Wong-Zakai theorem for stochastic PDEs. J. Math. Soc. Jpn. **67**, 1551–1604 (2015) (special issue for Kiyosi Itô.)
143. Hairer, M., Quastel, J.: A class of growth models rescaling to KPZ. arXiv:1512.07845
144. Hairer, M., Shen, H.: A central limit theorem for the KPZ equation. arXiv:1507.01237

145. Hofmanová, M., Röger, M., von Renesse, M.: Weak solutions for a stochastic mean curvature flow of two-dimensional graphs. Probab. Theory Relat. Fields (2016, published online)
146. Hohenberg, P.C., Halperin, B.I.: Theory of dynamic critical phenomena. Rev. Mod. Phys. **49**, 435–475 (1977)
147. Hora, A.: A diffusive limit for the profiles of random Young diagrams by way of free probability. Publ. RIMS Kyoto Univ. **51**, 691–708 (2015)
148. Hoshino, M.: KPZ equation with fractional derivatives of white noise. Stoch. Partial Differ. Equ. Anal. Comput. (2016, published online)
149. Hoshino, M.: Paracontrolled calculus and Funaki-Quastel approximation for KPZ equation. arXiv:1605.02624v2
150. Huisken, G.: Flow by mean curvature of convex surfaces into spheres. J. Differ. Geom. **20**, 237–266 (1984)
151. Huisken, G.: The volume preserving mean curvature flow. J. Reine Angew. Math. **382**, 35–48 (1987)
152. Huisken, G.: Asymptotic behavior for singularities of the mean curvature flow. J. Differ. Geom. **31**, 285–299 (1990)
153. Ilmanen, T.: Convergence of the Allen-Cahn equation to Brakke's motion by mean curvature. J. Differ. Geom. **38**, 417–461 (1993)
154. Ioffe, D.: Large deviations for the 2D Ising model: a lower bound without cluster expansions. J. Stat. Phys. **74**, 411–432 (1994)
155. Ioffe, D.: Exact large deviation bounds up to T_c for the Ising model in two dimensions. Probab. Theory Relat. Fields **102**, 313–330 (1995)
156. Ioffe, D., Schonmann, R.H.: Dobrushin-Kotecký-Shlosman theorem up to the critical temperature. Commun. Math. Phys. **199**, 117–167 (1998)
157. Johansson, K.: Shape fluctuations and random matrices. Commun. Math. Phys. **209**, 437–476 (2000)
158. Kaimanovich, V.A.: Dirichlet norms, capacities and generalized isoperimetric inequalities for Markov operators. Potential Anal. **1**, 61–82 (1992)
159. Kardar, M., Parisi, G., Zhang, Y.-C.: Dynamic scaling of growing interfaces. Phys. Rev. Lett. **56**, 889–892 (1986)
160. Katsoulakis, M.A., Kossioris, G.T., Lakkis, O.: Noise regularization and computations for the 1-dimensional stochastic Allen-Cahn problem. Interfaces Free Bound. **9**, 1–30 (2007)
161. Kawasaki, K.: Non-equilibrium and Phase Transition–Statistical Physics in Mesoscopic Scale, in Japanese. Asakura (2000)
162. Kawasaki, K., Ohta, T.: Kinetic drumhead model of interface I. Prog. Theoret. Phys. **67**, 147–163 (1982)
163. Kawasaki, K., Ohta, T.: Kinetic drumhead models of interface. II. Prog. Theoret. Phys. **68**, 129–147 (1982)
164. Kenyon, R.: Height fluctuations in the honeycomb dimer model. Commun. Math. Phys. **281**, 675–709 (2008)
165. Kenyon, R.: Lectures on dimers. In: Statistical Mechanics, pp. 191–230. I AS/Park City Mathematics Series, vol. 16. American Mathematical Society, Providence (2009)
166. Kenyon, R., Okounkov, A., Sheffield, S.: Dimers and amoebae. Ann. Math. **163**, 1019–1056 (2006)
167. Kerov, S.V.: Asymptotic representation theory of the symmetric group and its applications in analysis. Translations of Mathematics Monographs, vol. 219. American Mathematics Society, Providence (2003)
168. Kipnis, C., Landim, C.: Scaling Limits of Interacting Particle Systems. Springer, New York (1999)
169. Kipnis, C., Olla, S., Varadhan, S.R.S.: Hydrodynamics and large deviation for simple exclusion processes. Commun. Pure Appl. Math. **42**, 115–137 (1989)
170. Komorowski, T., Landim, C., Olla, S.: Fluctuations in Markov Processes: Time Symmetry and Martingale Approximation. Springer, Berlin/New York (2012)

171. Komorowski, T., Peszat, S., Szarek, T.: On ergodicity of some Markov processes. Ann. Probab. **38**, 1401–1443 (2010)
172. Kunita, H.: Stochastic Flows and Stochastic Differential Equations. Cambridge University Press, Cambridge/New York (1990)
173. Lacoin, H.: The scaling limit of polymer pinning dynamics and a one dimensional Stefan freezing problem. Commun. Math. Phys. **331**, 21–66 (2014)
174. Landim, C., Yau, H.-T.: Large deviations of interacting particle systems in infinite volume. Commun. Pure Appl. Math. **48**, 339–379 (1995)
175. Lee, K.: Generation and motion of interfaces in one-dimensional stochastic Allen-Cahn equation. arXiv:1511.05727
176. Lee, K.: Generation of interfaces for multi-dimensional stochastic Allen-Cahn equation with a noise smooth in space. arXiv:1604.06535
177. Lions, P.L., Souganidis, P.E.: Fully nonlinear stochastic partial differential equations. C. R. Acad. Sci. Paris Ser. I Math. **326**, 1085–1092 (1998)
178. Lions, P.L., Souganidis, P.E.: Fully nonlinear stochastic partial differential equations: non-smooth equations and applications. C. R. Acad. Sci. Paris Ser. I Math. **327**, 735–741 (1998)
179. Major, P.: Multiple Wiener-Itô Integrals, with Applications to Limit Theorems, 2nd edn. Lecture Notes Mathematics, vol. 849. Springer, Cham (2014)
180. Matano, H., Nakamura, K., Lou, B.: Periodic traveling waves in a two-dimensional cylinder with saw-toothed boundary and their homogenization limit. Netw. Heterog. Media **1**, 537–568 (2006)
181. Miller, J.: Fluctuations for the Ginzburg-Landau $\nabla \phi$ interface model on a bounded domain. Commun. Math. Phys. **308**, 591–639 (2011)
182. Mogul'skii, A.A.: Large deviations for trajectories of multi-dimensional random walks. Theory Probab. Appl. **21**, 300–315 (1976)
183. Mourrat, J.-C., Weber, H.: Convergence of the two-dimensional dynamic Ising-Kac model to Φ_2^4. Commun. Pure Appl. Math. (2016, published online)
184. Mourrat, J.-C., Weber, H.: Global well-posedness of the dynamic Φ^4 model in the plane. Ann. Probab. (2016, published online)
185. Mourrat, J.-C., Weber, H.: Global well-posedness of the dynamic Φ_3^4 model on the torus. arXiv:1601.01234
186. de Mottoni, P., Schatzman, M.: Geometrical evolution of developed interfaces. Trans. Am. Math. Soc. **347**, 1533–1589 (1995)
187. Mueller, C.: On the support of solutions to the heat equation with noise. Stoch. Stoch. Rep. **37**, 225–245 (1991)
188. Naddaf, A., Spencer, T.: On homogenization and scaling limit of some gradient perturbations of a massless free field. Commun. Math. Phys. **183**, 55–84 (1997)
189. Nagahata, Y.: A remark on equivalence of ensembles for surface diffusion model. RIMS Kôkyûroku Bessatsu **B59**, 23–30 (2016)
190. Nagahata, Y.: Spectral gap for surface diffusion (2015, preprint)
191. Nualart, D., Pardoux, E.: White noise driven quasilinear SPDEs with reflection. Probab. Theory Relat. Fields **93**, 77–89 (1992)
192. Petrov, V.V.: Sums of Independent Random Variables. Springer, Berlin/New York (1975)
193. Pimpinelli, A., Villain, J.: Physics of Crystal Growth. Cambridge University Press, Cambridge/New York (1998)
194. Pittel, B.: On a likely shape of the random Ferrers diagram. Adv. Appl. Math. **18**, 432–488 (1997)
195. Quastel, J.: Introduction to KPZ. In: Current Developments in Mathematics, vol. 2011, pp. 125–194. International Press, Somerville (2012)
196. Röger, M., Weber, H.: Tightness for a stochastic Allen-Cahn equation. Stoch. Partial Differ. Equ. Anal. Comput. **1**, 175–203 (2013)
197. Rybko, A., Shlosman, S., Vladimirov, A.: Spontaneous resonances and the coherent states of the queuing networks. J. Stat. Phys. **134**, 67–104 (2009)
198. Sasamoto, T., Spohn, H.: Superdiffusivity of the 1D lattice Kardar-Parisi-Zhang equation. J. Stat. Phys. **137**, 917–935 (2009)

199. Sasamoto, T., Spohn, H.: Exact height distributions for the KPZ equation with narrow wedge initial condition. Nuclear Phys. B **834**, 523–542 (2010)
200. Sasamoto, T., Spohn, H.: One-dimensional Kardar-Parisi-Zhang equation: an exact solution and its universality. Phys. Rev. Lett. **104**, 230602 (2010)
201. Shiga, T.: Two contrasting properties of solutions for one-dimensional stochastic partial differential equations. Canad. J. Math. **46**, 415–437 (1994)
202. Soner, H.M.: Ginzburg-Landau equation and motion by mean curvature. I. Convergence. J. Geom. Anal. **7**, 437–475 (1997)
203. Soner, H.M.: Ginzburg-Landau equation and motion by mean curvature. II. Development of the initial interface. J. Geom. Anal. **7**, 477–491 (1997)
204. Souganidis, P.E., Yip, N.K.: Uniqueness of motion by mean curvature perturbed by stochastic noise. Ann. Inst. H. Poincaré Anal. Non Linéaire **21**, 1–23 (2004)
205. Spohn, H.: Interface motion in models with stochastic dynamics. J. Stat. Phys. **71**, 1081–1132 (1993)
206. Spohn, H.: Nonlinear fluctuating hydrodynamics for anharmonic chains. J. Stat. Phys. **154**, 1191–1227 (2014)
207. Spohn, H., Stoltz, G.: Nonlinear fluctuating hydrodynamics in one dimension: the case of two conserved fields. J. Stat. Phys. **160**, 861–884 (2015)
208. Taylor, J., Cahn, J.W., Handwerker, C.A.: I-geometric models of crystal growth. Acta Metall. Meter. **40**, 1443–1474 (1992)
209. van Saarloos, W., Hohenberg, P.C.: Fronts, pulses, sources and sinks in generalized complex Ginzburg-Landau equations. Phys. D **56**, 303–367 (1992)
210. van Saarloos, W.: Front propagation into unstable states. Phys. Rep. **386**, 29–222 (2003)
211. Varadhan, S.R.S.: Nonlinear diffusion limit for a system with nearest neighbor interactions – II. In: Elworthy and Ikeda (eds.) Asymptotic Problems in Probability Theory: Stochastic Models and Diffusions on Fractals, pp. 75–128. Longman, Essex (1993)
212. Vershik, A.: Statistical mechanics of combinatorial partitions and their limit shapes. Funct. Anal. Appl. **30**, 90–105 (1996)
213. Vershik, A., Yakubovich, Yu.: The limit shape and fluctuations of random partitions of naturals with fixed number of summands. Mosc. Math. J. **1**, 457–468 (2001)
214. Vershik, A., Yakubovich, Yu.: Fluctuations of the maximal particle energy of the quantum ideal gas and random partitions. Commun. Math. Phys. **261**, 759–769 (2006)
215. Weber, H.: Sharp interface limit for invariant measures of the stochastic Allen-Cahn equation, Commun. Pure Appl. Math., **63** (2010), 1071–1109.
216. Weber, H.: On the short time asymptotic of the stochastic Allen-Cahn equation. Ann. Inst. H. Poincaré Probab. Statist. **46**, 965–975 (2010)
217. Weber, S.: The sharp interface limit of the stochastic Allen-Cahn equation. PhD thesis, University of Warwick (2014)
218. Weiss, G.S.: A free boundary problem for non-radial-symmetric quasi-linear elliptic equations. Adv. Math. Sci. Appl. **5**, 497–555 (1995)
219. Wilson, D.B.: Mixing times of lozenge tiling and card shuffling Markov chains. Ann. Appl. Probab. **14**, 274–325 (2004)
220. Wulff, G.: Zur Frage der Geschwindigkeit des Wachsthums und der Auflösung der Krystallflächen. Z. Krystallogr. **34**, 449–530 (1901)
221. Yakubovich, Yu.: Central limit theorem for random strict partitions. J. Math. Sci. **107**, 4296–4304 (2001)
222. Yip, N.K.: Stochastic motion by mean curvature. Arch. Ration. Mech. Anal. **144**, 313–355 (1998)
223. Yip, N.K.: Stochastic curvature driven flows. In: Da Prato, G., Tubaro, L. (eds.) Stochastic Partial Differential Equations and Applications. Lecture Notes in Pure and Applied Mathematics, vol. 227, pp. 443–460. Marcel Dekker, New York (2002)
224. Zhu, R., Zhu, X.: Three-dimensional Navier-Stokes equations driven by space-time white noise. J. Differ. Equ. **259**, 4443–4508 (2015)

Index

Q-Brownian motion, 87
$\frac{1}{3}$-power law, 113
$\nabla\varphi$-interface model, 8

A

Airy process, 70
Allen-Cahn equation, 83
anisotropic motion by mean curvature, 106
annihilation of box, 74
area of Young diagram, 30
asymptotic strong Feller property, 91
average tilt, 121

B

balance condition, 5, 24, 95
Bernoulli measure, 38, 66
Besov space, 123
bipartite planar graph, 71
bistable, 93
Boltzmann-Gibbs principle, 118
Bose statistics, 30
Burkholder-Davis-Gundy's inequality, 87

C

Cahn-Hilliard equation, 67, 83
canonical ensemble, 31, 68
capacity, 25
capacity bound, 26
capacity order, 19
carré du champ, 57
cave, 74
centering condition, 97

coexistence, 5, 14
Cole-Hopf solution, 115
Cole-Hopf transformation, 56, 114
Cole-Hopf transformation at microscopic level, 56
colored noise, 87
comparison theorem, 103, 114
concentration property, 8
convolution kernel, 115
correlation function, 75
coupled KPZ equation, 122
Cramér transform, 16
creation of box, 74
critical, 5, 6
current, 77
cylindrical Brownian motion, 86

D

decoupling estimate, 23
diagram formula, 119
diffusion coefficient, 79
diffusive scaling, 46, 76
dimer, 73
dimer configuration, 73
dimer cover, 73
dimer dynamics, 74
Dirichlet form, 66
Doeblin condition, 91
dry region, 24
dual lattice, 72
Duhamel's principle, 89
dynamic $P(\phi)_d$-model, 81
dynamic phase transition, 83

© The Author(s) 2016
T. Funaki, *Lectures on Random Interfaces*, SpringerBriefs in Probability
and Mathematical Statistics, DOI 10.1007/978-981-10-0849-8